PREFACE

This book offers an elementary approach to the theory of sets and symbolic logic and, in the final chapter, introduces the abstract system, Boolean algebra, to which these belong.

As you read the text, carefully study and work through the illustrative examples. You should then find yourself in a position to attempt the questions set at the end of the chapters.

Look critically at every statement made as part of a solution with a view to detecting and correcting inaccuracies, not only in working, but also in the use of symbols. A mathematical solecism such as $P \in Q$ is on a par with a grammatical error in an essay.

Following Chapters 6 and 10, some topics for investigation are suggested. These are intended to encourage interest in mathematical matters which may not in themselves have any special importance. Such investigation is valuable as a means of strengthening one's grasp of the subject; and the topics proposed here may serve to open the door upon a fascinating field of enquiry.

Questions taken from examination papers are indicated thus:

AEB: Associated Examining Board for the GCE.

JMB: Joint Matriculation Board (Universities of Manchester, Liverpool, Leeds, Sheffield, Birmingham).

I gratefully acknowledge my indebtedness to my colleague, Miss Florence Berryman for her valuable comments and for checking the answers: also to the Publishers for their unfailing consideration and helpful co-operation.

Belfast
March 1970

H T COMBE

CONTENTS

1 | DEFINITIONS AND NOTATION

1.1 The idea of a set

A *set* is a well defined collection of objects. Two things about this description may appear strange.

(*i*) The concept of a set embraces objects of any kind, and not merely numbers. The relation between the new algebra of sets and the familiar algebra of numbers will become clearer as we proceed: the important consideration at present is to become thoroughly familiar with the idea of a set and other ideas connected with it.

(*ii*) Also, it may be remarked that in the above description of a set a number of more difficult words are used to explain a very simple one. Despite this, it should be remembered that the idea of a set is just as simple as it appears to be. A collection of British postage stamps, the positive integers, a group of people in a room, the words on this page, are all examples of sets.

The words 'well defined' should be noticed. They express the fact that there should be no doubt as to whether a given object belongs, or does not belong, to the set under consideration.

Example 1. If A_4 is the set of four-legged animals, a mallard does not belong to A_4 since, though it is an animal, it does not have four legs. Nor does a table belong to A_4 since, though it may have four legs, it is not an animal. A_4 is a well defined collection of objects.

Example 2. If Z^{E+} is the set of even positive integers, it is immediately clear that $8\cdot5$, -4, 7 are all excluded from Z^{E+}: why?

Example 3. If S is the set of solutions of the equation $x^2 - 5x + 4 = 0$, then the numbers 1 and 4 belong to S, and no others do. S is called the *solution set* of the equation $x^2 - 5x + 4 = 0$. We say that 1 and 4 are the *elements* of S.

Notation

The phrase 'is an element of' or 'belongs to' is denoted by $\in$. In Example 3, $1 \in S$, $4 \in S$.

Since, when $x = 2$, $x^2 - 5x + 4 = -2$ (and not 0), 2 is not an element of S: we write $2 \notin S$. A vertical or oblique stroke through any sign denotes the negation of the statement represented by that sign. What do $\neq$, $\not>$, $\not<$ mean? What is the difference in meaning between $a \not> b$ and $a < b$?

It will assist in gaining familiarity with a new notation if it is translated into words when it is being read. This applies generally to all the new mathematical symbols which will be introduced in this book.

Exercise 1a

By replacing ? by either $\in$ or $\notin$, make a true statement of each of the following:

1. If A is the set of essential components of a car engine, a windscreen wiper ? A.
2. If B is the set of people forming a beat group, a drummer ? B.
3. If C is the solution set of the equation $x = \dfrac{1}{x}$, 0 ? C.
4. If D is the set of sources of energy, the moon ? D.
5. If E is the set of numbers which are perfect squares, $2\frac{1}{4}$? E.

1.2   Set notation

The number of elements in a set may be finite or infinite. Thus, if V is the set of vowels in the English language, there are five and only five elements of V. But if N is the set of positive integers, N has an infinite number of elements.

There are two usual ways of writing a given set.

The roster notation

Where the number of elements is small they may be listed (written in a roster), in which case they are enclosed in curly brackets. Thus the set V of vowels in the English alphabet may be written:

$$V = \{a, e, i, o, u\}$$

Attention should be given to three points about this notation.

(*i*) The order in which the elements are written does not matter. We may equally well write $V = \{i, u, e, a, o\}$.

(*ii*) No element is repeated. Thus if W is the set of vowels in the word *parachute*, then $W = \{a, e, u\}$, even although a appears twice in the word.

(*iii*) If the number of elements is large, some only need be written, the rest being represented by dots. This is possible only where there is no doubt as to the identity of the omitted elements. Thus the set $E = \{2, 4, 6, \ldots 20\}$ is understood as the set of even positive integers not greater than 20.

A set may have only one element. For example, the solution set S of the equation $3x - 1 = 20$ is $\{7\}$. There is a logical distinction between the set $\{7\}$ and the number 7: in fact, $7 \in \{7\}$. Although it does not matter whether we say 'The solution set of the equation $3x - 1 = 20$ is $\{7\}$', or say, 'If $3x - 1 = 20$, then $x = 7$', the former statement has the advantage of bringing the notation for simple equations into line with that for equations of higher degree.

The set-builder notation

Where the roster notation is unwieldy or impracticable, we give a description of the elements of the set, taking care in the description to define the set completely.

Example 4. If M is the set of men who sit in the House of Commons, we may write:

$$M = \{x \mid x \text{ is a man, } x \text{ is an MP}\}$$

The line | (sometimes replaced by :) stands for 'such that'; following it is a complete, unambiguous definition of x which enables the elements x of M to be recognized. In this example, x must satisfy two conditions.

Example 5. $K = \{n \mid n \text{ is an integer, } 9 < n < 100\}$ is read 'K is the set of all n such that n is an integer greater than 9 and less than 100': or, more simply, 'K is the set of all two-digit positive integers'.

When a set, defined in the set-builder notation, is being described in words, a form of words should be chosen that is as concise as possible consistent with accuracy.

Note that the set K could be written $K = \{10, 11, 12, 13, \ldots 99\}$, since it is clear what elements are represented by the dots.

Exercise 1b

1. Write the sets A, B, C, D, E in roster notation:

    A is the set of months of the year having not more than 30 days.

    B is the set of single-digit positive integers not divisible by 3.

    C is the set of those capital letters of the English alphabet which can be written on transparent paper so as to appear unchanged when inverted.

    D is the set of all triangles in the parallelogram $PQRS$ whose diagonals intersect at O.

    $E = \{n \mid n$ is a positive even number, $3n < 32\}$.

2. Write the sets F, G, H, K, L in set-builder notation:

    $F = \{0, 1, 4, 9, 16, 25, 36\}$

    $G = \{2, 5, 8, 11, 14, \ldots\}$

    $H = \{$North, South, East, West$\}$

    $K = \{1, \sqrt{2}, \sqrt{3}, 2, \sqrt{5}, \sqrt{6}, \sqrt{7}, \sqrt{8}, 3\}$

    $L = \{1, \frac{1}{2}, \frac{1}{3}, \frac{1}{4}, \ldots \frac{1}{20}\}$

3. Describe in words the sets X, Y, Z:

    $X = \{1, 2, 4, 8, 16, 32, \ldots\}$

    $Y = \{x \mid x = 5y, y$ is an integer$\}$

    $Z = \{$physics, chemistry, biology$\}$

1.3 Equality of sets

Two sets X and Y are *equal* if and only if they have the same elements. We then write $X = Y$.

This definition means that two tests have to be applied for equality. We have to ask:

   (i) Is it true that if $x \in X$, then $x \in Y$?

   (ii) Is it true that if $y \in Y$, then $y \in X$?

Example 6. If $A = \{1, 8, 27, 64\}$, $B = \{x \mid x = y^3, y$ is a positive integer, $y \not> 4\}$, then $A = B$.

Example 7. If $P = \{4, 6, 8\}$, $Q = \{0, 2, 4, 6, 8, 10\}$, P and Q have the elements 4, 6, 8, in common, and every element of P is an element of Q; but not every element of Q is an element of P. Hence $P \neq Q$.

Example 8. If X = {He, Ne, Ar, Kr}, Y = {Ne, Ar, Kr, Xe}, then $X \neq Y$ since He $\in X$, but He $\notin Y$; and Xe $\in Y$, but Xe $\notin X$.

Exercise 1c

In each of the following cases, state whether $A = B$ or $A \neq B$:
1. A = {2, 1·8, 1·6, 1·4, 1·2, 1}
   B = {1, $1\frac{1}{5}$, $1\frac{2}{5}$, $1\frac{3}{5}$, $1\frac{4}{5}$, 2}
2. A = {$x \mid x$ is a solution of the equation $x^2 - x = 0$}
   B = {$x \mid x$ is a solution of the equation $x - 1 = 0$}
3. A = {$x \mid x = 1 + \dfrac{1}{y}$, y is a positive integer, $y < 7$}
   B = {2, $1\frac{1}{2}$, $1\frac{1}{3}$, $1\frac{1}{4}$, $1\frac{1}{5}$, $1\frac{1}{6}$}
4. A = {$x \mid x$ is a player's position in a rugger side}
   B = {scrum half, left wing, right wing, full back}

1.4 Subsets

If X and Y are two sets, and if X consists of elements all of which are elements of Y, X is called a *subset* of Y.

Let X = {1, 3, 5, 7}, Y = {$x \mid x$ is a positive integer, $x < 10$}. Then since all the elements of X are elements of Y, X is a subset of Y.

Notation

'X is a subset of Y' is written $X \subset Y$. The symbol $\subset$ may be read either as 'is a subset of' or as 'is contained in'.

From the definition, any set Y is a subset of itself: we say 'Y is an *improper* subset of Y'. Since $Y = Y$, the notation $X \subseteq Y$ is used to include both proper and improper subsets X of Y. $X \subseteq Y$ is read 'X is contained in, or equals, Y'.

If X is a subset of Y, Y is sometimes called a *superset* of X. This is written $Y \supset X$, and may be read as 'Y contains X'. $Y \supseteq X$ means 'Y contains, or is equal to, X'.

Note. The difference in meaning between the symbols $\in$ and $\subset$ should be carefully observed. The former expresses a relation between an element and a set: the latter, a relation between two sets.

Theorem I

If A, B, C are three sets and $A \subset B$, $B \subset C$, then $A \subset C$.

The theorem may be established by a *direct* proof. This consists of a series of steps (deductions) by which, using the data, the required conclusion is reached. A proof is valid if and only if each step follows logically from another. This subject will be developed in Chapter 9.

Proof. Let $a \in A$, then $a \in B$ (since $A \subset B$). If $a \in B$, then $a \in C$ (since $B \subset C$). So, if $a \in A$, then $a \in C$, *i.e.*, $A \subset C$.

Exercise 1d

1. Let P be the set of candidates who sat for an examination, Q be the set of those who scored over 80%, R be the set of those who scored over 55%, S be the set of those who gained a pass-mark of 40% or more, and T be the set of failures.

   In each of the following, replace ? by an appropriate symbol to make it a true statement:

   (a) $T \,?\, P$ (b) $Q \,?\, R$ (c) $R \,?\, Q$ (d) $Q \,?\, S$ (e) $S \,?\, R$

2. Sets A, B, C, D are defined as follows:

   $A = \{x \mid x$ is a positive integer, $x < 50\}$

   $B = \{x \mid x = 2y$, y is a positive integer, $y < 11\}$

   $C = \{x \mid x$ is an odd positive integer, $x < 100\}$

   $D = \{x \mid x = y^2$, y is a positive integer, $y < 8\}$

   Name all the sets which are subsets of A. Is any set a superset of A?

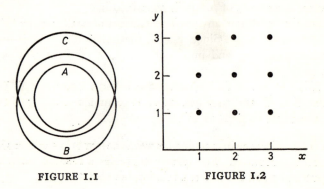

FIGURE I.I   FIGURE I.2

3. In Fig. 1.1, if A, B, C are the sets of points within the three circles, state a relation between :
(a) A and B (b) C and A (c) B and A
Make a statement, in words and in symbols, about B and C.

4. Let L be the set of points (the *lattice*) shown in Fig. 1.2: then
$L = \{(1, 1), (1, 2), (1, 3), (2, 1), (2, 2), (2, 3), (3, 1), (3, 2), (3, 3)\}$
If $A = \{(1, 2), (2, 2), (3, 2)\}$, $B = \{(1, 2), (1, 3), (1, 4)\}$, and
$C = \{(1, 2), (1, 3), (2, 2), (3, 1), (3, 2)\}$, state a relation between:
(a) A and C (b) C and L (c) B and L (d) L and A

1.5 The empty (or null) set

Consider the set X defined by:
$$X = \{x \mid x \text{ is a multiple of } 7,\ 8 < x < 12\}$$
There is clearly no multiple of 7 in the given interval, hence X has no elements.

The set which has no elements is called the *empty* (or *null*) set.

Notation

The empty set is denoted by $\emptyset$ or, sometimes, by the symbol $\{\}$.

The set $\emptyset$ should be carefully distinguished from the set $A = \{0\}$. A has one element, 0: $\emptyset$ has no element.

Since if X is any set, $\emptyset$ does not contain any element which does not belong to X, it follows that $\emptyset \subset X$. Thus, the empty set is a subset of every set.

Example 9. The subsets of $\{x_1, x_2\}$ are $\{x_1, x_2\}$, $\{x_1\}$, $\{x_2\}$, $\emptyset$.

We may speak of 'the set S_2 of subsets of $\{x_1, x_2\}$'. There are four elements of S_2, each element being itself a set. Thus:
$$S_2 = \{\{x_1, x_2\}, \{x_1\}, \{x_2\}, \emptyset\}$$
S_2 is called the *power set* of $\{x_1, x_2\}$.

It need not be thought strange that the elements of a set may be sets. A set may consist of any objects, and these may be sets.

1.6 The universal set

In Example 8 (p. 5), X and Y are sets of chemical elements, helium, neon, argon, *etc.* There is a complete set of the chemical elements at present known, of which X and Y are subsets. This complete set is the *universal set* for this example.

Notation

In any given case, the universal set, *i.e.*, the complete set to which all elements under consideration belong, is denoted by $\mathscr{E}$, or, sometimes, by U.

While there is only one empty set $\emptyset$, the universal set $\mathscr{E}$ depends on the nature of the elements being considered. In Example 6 (p. 4), for instance, $\mathscr{E}$ would be the set of cubes of all the positive integers. If X is the set of letters in the word *azimuth*, $\mathscr{E}$ is the set of letters in the English alphabet. In some cases, where it cannot be inferred, $\mathscr{E}$ is specifically defined.

Note that for every universal set $\mathscr{E}$, $\emptyset \subset \mathscr{E}$.

1.7 Illustrating sets

It is often helpful to draw a diagram illustrating given sets and the relations between them. The usual type of diagram is shown in Fig. 1.3.

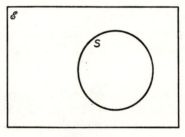

FIGURE I.3

The rectangle represents the universal set, and the circle a given set S. The placing of the circle entirely within the rectangle illustrates the relation $S \subset \mathscr{E}$.

The dimensions of the figures are irrelevant, and it is not necessary to draw exact geometrical shapes: any loop will serve to illustrate S, provided it is correctly placed in relation to the other shapes included. Such diagrams are called *Venn–Euler*, or, more commonly, *Venn* diagrams. (Euler was a Swiss mathematician who lived from 1707 to 1783; John Venn, 1834–1923, was an English mathematician and logician.)

Where, in any given example, $\mathscr{E}$ does not come into consideration, the rectangle need not be included in the diagram. Fig. 1.1 (p. 6) is another example of a Venn diagram.

Exercises on Chapter 1

1. Write the solution set of each of the following equations. In each case, x is a real number.

   (a) $13x + 2 = 54$ (e) $3x - 2 = (x + 1) + (2x - 3)$
   (b) $x(x + 11)(2x - 3) = 0$ (f) $1/(x + 2) = x + 2$
   (c) $4x^2 + 1 = 0$ (g) $(x - 2)(x - 3) = 2$
   (d) $5/x^2 = 45$

2. Write in roster notation the power set S_3 of the set $\{x_1, x_2, x_3\}$. How many elements has S_3 (cf. Example 9 on p. 7)? How many elements do you think there are in:

   (a) S_4, the power set of $\{x_1, x_2, x_3, x_4\}$
   (b) S_n, the power set of $\{x_1, x_2, x_3, \ldots, x_n\}$

3. If P is the set of all parallelograms, R is the set of all rectangles, S is the set of all squares, examine the truth of each of the following statements:

   (a) $S \subset R$ (b) $P \subset R$ (c) $S \subset P$

4. Draw a Venn diagram showing three sets A, B, C such that $A \subset C$, $C \subset B$.

5. Name a set which has only one subset.

6. Two sets P, Q are said to be *comparable* if $P \subset Q$ or $Q \subset P$: otherwise they are *non-comparable*. Draw two different Venn diagrams each showing two non-comparable sets P, Q.

7. If $A \subset C$ and $B \subset C$, are A and B possibly comparable? Draw Venn diagrams to support your conclusion. Are A and B necessarily comparable?

8. Let B be the set of all Scottish boys, C be the set of all children, A be the set of all Scottish people, and D be the set of all boys. Write out all the relations between A, B, C, D that can be expressed in terms of $\subset$. Which, if any, of these sets are non-comparable?

9. Write in roster notation the sets A, B, C, D, E:

   $A = \{f \mid f \text{ is a positive proper fraction with denominator 6}\}$

$B = \{x \mid x \text{ is a factor of } 72\}$
$C = \{m \mid m \text{ is a multiple of } 7, m < 50\}$
$D = \{s \mid s \text{ is one of the human senses}\}$
$E = \{d \mid d \text{ is a digit of the number } 5\,826\,569\}$

10. If $\mathscr{E}$ is the set of all real numbers, make a true statement about each of the following sets:
    (a) $\{n \mid n^2 > 0\}$ (c) $\{n \mid 0.n < 0\}$
    (b) $\{n \mid (2n-1)(2n+1) = 4n^2 - 1\}$ (d) $\{n \mid 2n \nless n^2 + 1\}$

11. If $P = \{1, 2, 3, 4, 7\}$, $Q = \{3, 4, 5, 6, 7, 8\}$, $R = \{3, 4, 7, 8\}$, write down all the possibilities for a set X such that $X \subset Q$, $X \subset R$, $X \not\subset P$.

12. Give a direct proof of the truth of the statement: If A and B are two sets such that $A \subseteq B$ and $B \subseteq A$, then $A = B$. (The first step might be the same as for Theorem I on p. 6.)

2 | OPERATIONS WITH SETS

2.1 Operations

When we write $18 + 3 = 21$, $18 \times 3 = 54$, $18 - 3 = 15$, we are performing *operations* on the two numbers 18 and 3. These operations assign to 18 and 3 numbers 21, 54 and 15 which we call the sum, product and positive difference of 18 and 3.

In the algebra of sets also there are operations. They have different names from those performed on numbers, and are usually denoted by different symbols. In some respects they are like the familiar operations on numbers: in others they differ from these operations.

2.2 Union

Definition

If X and Y are two sets, the set of elements which belong to X, or to Y, or to both X and Y, is called the *union* of X and Y.

Notation

The union of X and Y is written $X \cup Y$. Thus $X \cup Y = \{x \mid x \in X$ or $x \in Y\}$, where 'or' is understood in the sense of 'and/or'.

$X \cup Y$ is sometimes read as 'X cup Y', but this is more clever than helpful. Read $X \cup Y$ as 'X union Y' or 'X or Y'.

Example 1. If $P = \{3, 6, 9, 12\}$ and $Q = \{5, 10, 15\}$, then the union $P \cup Q = \{3, 5, 6, 9, 10, 12, 15\}$.

Example 2. If $X = \{1, 2, 3, 4\}$ and $Y = \{3, 4, 5\}$, then the union $X \cup Y = \{1, 2, 3, 4, 5\}$.

Example 3. If $A = \{x \mid x$ is an integer, $x > 5\}$ and $B = \{x \mid x$ is an integer, $x < 6\}$, then $A \cup B = \{x \mid x$ is an integer$\}$.

11

If in Fig. 2.1 the circles represent the sets X and Y, the shaded area represents the set $X \cup Y$. This Venn diagram illustrates Example 2.

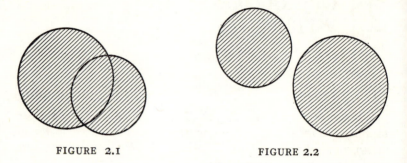

FIGURE 2.1   FIGURE 2.2

Figure 2.2 illustrates Example 1. Why is it drawn differently from Fig. 2.1?

In Example 3 the sets A and B together contain all the elements of the universal set of integers. An appropriate Venn diagram for $A \cup B$ in this case is shown in Fig. 2.3.

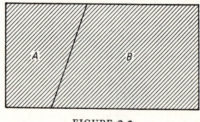

FIGURE 2.3

It is clear from the definition of union that $P \cup Q = Q \cup P$, since these two sets consist of the same elements. When, as in this case, it does not matter in what order the quantities being operated with are taken, the operation is said to be *commutative* for these quantities.

The union of sets is a commutative operation.

Since, if a, b are numbers, $a.b = b.a$, multiplication is a commutative operation on two numbers. Is addition commutative for numbers? Is subtraction?

Exercise 2a

1. Write down $P \cup Q$ in each of the following cases:
   (a) $P = \{-2, 0, 2, 4\}$, $Q = \{-4, 4\}$
   (b) $P = \{1, -1, 3\}$, $Q = \{2, -2, -3\}$
   (c) $P = \left\{\dfrac{a}{b}, \dfrac{b}{c}, \dfrac{c}{d}\right\}$, $Q = \left\{\dfrac{b}{c}\right\}$

2. If $A = \{\theta \mid \theta$ is a letter of the word *process*$\}$
   $B = \{\theta \mid \theta$ is a letter of the word *escort*$\}$
   write in roster notation the sets A, B, $A \cup B$.

3. Write in set-builder notation the set $X \cup Y$ in each of the following cases:
   (a) $X = \{\frac{1}{2}, \frac{1}{4}, \frac{1}{6}, \frac{1}{8}, \frac{1}{10}\}$, $Y = \{1, \frac{1}{3}, \frac{1}{5}, \frac{1}{7}, \frac{1}{9}\}$
   (b) $X = \{x \mid x = 2n, n$ is a positive integer, $n < 6\}$
   $Y = \{x \mid x = 4n, n$ is a positive integer, $n < 4\}$
   (c) $X = \{x \mid x$ is a solution of the equation $9x^2 - 15x + 4 = 0\}$
   $Y = \{x \mid x$ is a solution of the equation $3x^2 - 7x + 2 = 0\}$

4. By means of Venn diagrams illustrate the truth of the statements:
   (a) $A \cup \mathscr{E} = \mathscr{E}$
   (b) $A \subset (A \cup B)$
   (c) If $A \subset B$, then $A \cup B = B$

2.3 Intersection

Definition

If X and Y are two sets, the set of all elements which belong to both X and Y is called the *intersection* of X and Y.

Notation

The intersection of X and Y is written $X \cap Y$. Thus:

$$X \cap Y = \{x \mid x \in X \text{ and } x \in Y\}$$

$X \cap Y$ is sometimes read as 'X cap Y': again, it is more helpful to read it as 'X intersection Y' or 'X and Y'.

Example 4. If $X = \{1, 2, 3\}$, $Y = \{2, 3, 4\}$, then $X \cap Y = \{2, 3\}$.
   Draw a Venn diagram to represent X and Y, and shade the region which represents $X \cap Y$.

Example 5. If $\mathcal{J}$ is the set of all jet aircraft in operation, and A is the set of all aircraft operated by BOAC, then $\mathcal{J} \cap A$ is the set of BOAC jets.

Example 6. If P_1 is the set of all points on a tangent to a circle, and P_2 is the set of all points within, but not on, the circle, what can be said about $P_1 \cap P_2$?

Where two sets A, B have no elements in common, *i.e.*, where $A \cap B = \emptyset$, A and B are said to be *disjoint* sets. Which of the Venn diagrams in Figs. 2.1, 2.2 and 2.3 represent disjoint sets?

Example 7. Let:

$$A = \{x \mid x \text{ is a positive integer, } x < 8\}$$
$$B = \{x \mid x \text{ is an even positive integer, } x < 12\}$$
$$C = \{x \mid x \text{ is a positive integer, } 2 < x < 10\}$$

Write A, B, C in roster notation and check the following results:

(*i*) $A \cap B = \{2, 4, 6\}$ (*iv*) $(A \cap B) \cap C = \{4, 6\}$
(*ii*) $B \cap C = \{4, 6, 8\}$ (*v*) $A \cap (B \cap C) = \{4, 6\}$
(*iii*) $C \cap A = \{3, 4, 5, 6, 7\}$

Note that in this case $(A \cap B) \cap C = A \cap (B \cap C)$. This will be seen in a later chapter to be true in general for any three sets P, Q, R. We say intersection is an *associative* operation in P, Q, R, since in finding the intersection of P, Q, R it does not matter which pair is taken first. This being so, we may write the intersection of P, Q, R as $P \cap Q \cap R$.

In this respect, intersection of sets is analogous to multiplication of numbers, which is also associative, since $(a.b).c = a.(b.c)$ for all numbers a, b, c.

Example 8. Let A, B, C be the three sets defined in the previous example. Check the following results:

(*i*) $B \cup C = \{2, 3, 4, 5, 6, 7, 8, 9, 10\}$
(*ii*) $A \cap (B \cup C) = \{2, 3, 4, 5, 6, 7\}$
(*iii*) $A \cap B = \{2, 4, 6\}$
(*iv*) $A \cap C = \{3, 4, 5, 6, 7\}$
(*v*) $(A \cap B) \cup (A \cap C) = \{2, 3, 4, 5, 6, 7\}$

Compare the results in (*ii*) and (*v*): they show that in this case $A \cap (B \cup C) = (A \cap B) \cup (A \cap C)$. This also will be seen in a later chapter to be a general law, true for any three sets P, Q, R. The law states that in the algebra of sets, intersection *distributes* over union.

This distributive law also holds for any three numbers: multiplication distributes over addition, since $a.(b + c) = a.b + a.c$.

Exercise 2b

1. If $X = \{$He, Ne, Ar, Kr$\}$ and $Y = \{$Ne, Ar, Kr, Xe$\}$, write down the set $X \cap Y$.

2. A and B are two sets. With the help of Venn diagrams if necessary, complete the following to make true statements:
   (*a*) If $A = B$, then $A \cap B = ?$ and $A \cup B = ?$
   (*b*) If $A \subset B$, then $A \cap B = ?$
   (*c*) $A \cap \emptyset = ?$
   (*d*) $A \cap \mathscr{E} = ?$

3. Draw a Venn diagram to illustrate three sets X, Y, Z such that $X \subset Y$ and $Y \cap Z = \emptyset$.

4. We have seen that union is a commutative operation in two sets P, Q (p. 12). Is intersection also a commutative operation in P, Q?

5. We have seen that intersection is an associative operation in three sets P, Q, R. Is union also an associative operation in P, Q, R? Is addition an associative operation on three numbers a, b, c?

6. Let A, B, C be the three sets defined in Example 7. Write down the following sets:

   (*a*) $B \cap C$ (*d*) $A \cup (B \cap C)$
   (*b*) $A \cup B$ (*e*) $(A \cup B) \cap (A \cup C)$
   (*c*) $A \cup C$

   Compare the last two results: they should be the same. This illustrates the general law that for any three sets, union distributes over intersection.

7. Is it true that for any three numbers a, b, c addition distributes over multiplication: in other words, is $a + b.c = (a+b).(a+c)$ true?

Note. In question 7 it may easily be shown by taking values for a, b, c, that the relation $a + b.c = (a + b).(a + c)$ is not generally true. Thus, if $a = 1$, $b = 2$, $c = 3$, $a + b.c = 7$ but $(a + b).(a + c) = 12$. This single failure is sufficient to disprove the general statement, even if its falsity were not evident in any other way. Such an example, which disproves a general statement by showing it to be false in one instance, is called a *counter example*.

Observe carefully that a theorem cannot be proved true in general by quoting one special case in which it is true (or even a number of such cases). For example, check that the following table gives four sets of values of a, b, c for all of which $a + b.c = (a + b)(a + c)$:

a	1	2	3	4
b	-2	3	-1	0
c	2	-4	-1	-3

Yet the relation does not hold in general since a case of failure has been found.

A theorem cannot be proved true by quoting examples in which it holds. It can, however, be proved false by quoting one counter example.

2.4 Complement

Let the universal set $\mathscr{E}$ be the total labour force in a given factory, and let P be the set of workers who are present on a given day. Then the set of those who are absent is called the *complement* of P in $\mathscr{E}$.

Definition

The complement of a given set P is the set of those elements of $\mathscr{E}$ which do not belong to P.

Notation

The complement of P is written P'. Copy Fig. 1.3 (p. 8) and shade the region which represents S'.

Example 9. If $\mathscr{E}$ is the set of letters in the English alphabet and V is the set of vowels, then V' is the set of consonants.

Example 10. If Z is the set of integers, and Z^+ is the set of positive integers, then $(Z^+)'$ is the set of negative integers and zero.

Exercise 2c

1. If $\mathscr{E} = \{x \mid x$ is a single-digit positive integer$\}$, $A = \{1, 3, 7, 9\}$, $B = \{1, 5, 9\}$, $C = \{2, 4, 8\}$, write out the sets:

(a) A' (e) $A' \cup C'$
(b) B' (f) $A' \cap C'$
(c) C' (g) $A' \cap B' \cap C'$
(d) $A' \cap C$

2. If P, Q, R are three sets illustrated in Fig. 2.4, draw Venn diagrams carefully shaded or coloured to show:

(a) $Q' \cup R$ (c) $(P \cup Q \cup R)'$
(b) $(P \cap Q \cap R)'$

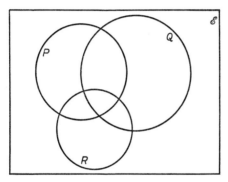

FIGURE 2.4

3. By quoting a counter example, prove that the statement $(A \cap B)' = A' \cap B'$ is false.

2.5 Relative complement or difference

Definition

If X and Y are two sets, the set of elements which belong to X but not to Y is called the *relative complement* or *difference set*.

Notation

The complement of X relative to Y is denoted by $X - Y$. This should be read 'X difference Y' and not 'X minus Y'.

Similarly, the complement of Y relative to X is $Y - X$.

Example 11. If $A = \{0, 1, 2, 3, 4, 5, 6\}$ and $B = \{0, 2, 4, 6, 8\}$, then $A - B = \{1, 3, 5\}$, and $B - A = \{8\}$.

In Fig. 2.5, the area representing $A - B$ is shaded. Copy the figure and shade the area representing $B - A$. Study your diagram, and make a statement about the set $(A - B) \cup (A \cap B) \cup (B - A)$.

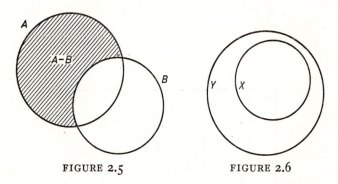

FIGURE 2.5   FIGURE 2.6

Example 12. Examine the truth of the statement: 'If X and Y are two sets, and $X \subset Y$, then $X - Y = \emptyset$.'

If $x \in X$ and $X \subset Y$, then $x \in Y$ (definition of a subset). This is true for all $x \in X$: there are no elements of X which do not belong to Y, so $X - Y = \emptyset$. The Venn diagram is shown in Fig. 2.6.

Exercise 2d

1. If $\mathscr{E} = \{x \,|\, x \text{ is a single-digit positive integer}\}$, $A = \{1, 3, 7, 9\}$, $B = \{1, 5, 9\}$, $C = \{2, 4, 8\}$, write out the sets:
   (a) $A - B$ (d) $B - A'$
   (b) $A' - C$ (e) $A - \emptyset$
   (c) $A' - B'$ (f) $\emptyset - A$
2. Show by means of a counter example that the statement 'Set difference is a commutative operation' is false.

3. Copy Fig. 2.7 and shade the area which represents the set $A \cap B'$. Compare the result with Fig. 2.5. What conclusion do you draw? Make a corresponding statement about $B \cap A'$.

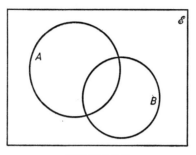

FIGURE 2.7

2.6 Symmetric difference

Definition

If P and Q are two sets, the set $(P \cup Q) - (P \cap Q)$ is called the *symmetric difference* of P and Q.

Notation

The symmetric difference of P and Q is denoted by $P \, \Delta \, Q$. Thus:
$$P \, \Delta \, Q = (P \cup Q) - (P \cap Q)$$

Example 13. If $\mathscr{E} = \{x \mid x$ is a single-digit positive integer$\}$, and if $A = \{1, 2, 3, 4, 5, 6\}$ and $B = \{2, 4, 6, 8\}$, then $A \, \Delta \, B = \{1, 3, 5, 8\}$, $A \, \Delta \, \emptyset = A$, $A \, \Delta \, \mathscr{E} = A' = \{7, 8, 9\}$. Check these results.

Example 14. If S_1 is the set of pupils in a form who take French and German and S_2 is the set of those who take French and Latin, then $S_1 \, \Delta \, S_2$ is the set of those who take Latin and German.

In Fig. 2.8, $P \cup Q$ is shaded horizontally and $P \cap Q$ vertically. Copy the figure and shade obliquely the regions which represent $P \, \Delta \, Q$. Compare the result with Fig. 2.5 (p. 18) and the Venn diagram drawn for Example 11 (p. 18). Deduce an alternative definition of $P \, \Delta \, Q$.

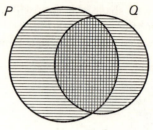

FIGURE 2.8

Exercise 2e

1. If $\mathscr{E} = \{x \mid x$ is a single digit positive integer$\}$, $A = \{1, 3, 7, 9\}$, $B = \{1, 5, 9\}$, $C = \{2, 4, 8\}$, write out the sets:
   (a) $A \, \Delta \, B$ (d) $(A \cap C') \, \Delta \, B$
   (b) $A' \, \Delta \, C$ (e) $(A \cup B) \, \Delta \, C'$
   (c) $(A \cap B) \, \Delta \, C$ (f) $(A - B) \cup (B - A)$
2. Let $A = \{p, q, r, s, t, u\}$, $B = \{s, t, u, v, w\}$, $C = \{q, r, v\}$. Write out the sets:
   (a) $(A \, \Delta \, B) - C$ (c) $(A - B) \, \Delta \, C$
   (b) $A - (B \, \Delta \, C)$ (d) $A \, \Delta \, (B - C)$
3. Copy Fig. 2.4 (p. 17) and carefully shade the regions which represent $P \, \Delta \, (Q \, \Delta \, R)$. What does the result suggest as to the relation between $P \, \Delta \, (Q \, \Delta \, R)$ and $(P \, \Delta \, Q) \, \Delta \, R$? Does it seem justifiable to write these simply as $P \, \Delta \, Q \, \Delta \, R$?

Care should be taken to distinguish between *operations* and *relations*. Each of the symbols $=$, $\subset$ and $\supset$ used with two sets A, B denotes a relation between A and B: it makes a *statement* about A and B.

On the other hand, the symbols $\cup$, $\cap$, $-$ and Δ used with two sets A, B define new sets obtained by operating in A and B. The same is true of the symbol $'$ since $A' = \mathscr{E} - A$.

As an illustration of the difference between a relation and an operation, refer to Example 7 (p. 14) where it was found that $(A \cap B) \cap C = A \cap (B \cap C)$. This states a relation (that of equality) between two sets: but each of these sets has been obtained by a repeated operation ($\cap$) in A, B, C.

Note. The symbol $\Rightarrow$ denotes *logically implies*. It may appropriately be used when a deduction is made with a valid reason. Thus the statement $x \in A$ and $x \in B \Rightarrow x \in (A \cap B)$ is true for all sets A, B from the definition of intersection.

Exercises on Chapter 2

1. Let $\mathscr{E} = \{n \mid n$ is a positive integer, $n < 20\}$, $A = \{n \mid n$ is a positive integer, $4 < n < 16\}$, $B = \{n \mid n$ is a positive integer, $10 < n < 20\}$, $C = \{n \mid n$ is a positive integer, $5 < n < 12\}$. Write out the sets:
   (a) $(A' \triangle B) \cap (B' \triangle C)$ (c) $(A' - B') - C'$
   (b) $(A' \cap B) \triangle (B \cap C')$ (d) $C' - (A' - B')$

2. If $\mathscr{E}$ is the set of words in the Oxford Dictionary, A is the set of monosyllables, B is the set of words beginning with f, C is the set of words ending in *-tion*:
   (i) state the meaning of the sets:
       (a) $A \cap B$ (d) $A \triangle C$
       (b) $A - B$ (e) $A' \cap B \cap C$
       (c) $B - (A \cup C)$
   (ii) name (in terms of A, B, C) three sets to which the word *fruition* belongs, and three to which *amplitude* belongs
   (iii) make a statement about the set $A \cap C$

3. If D_{24} is the set of all factors of 24 (including 1 and 24) and D_{54} is the set of all factors of 54 (including 1 and 54), write out D_{24} and D_{54}.
   Write out the set $D_{24} \cap D_{54}$ and explain its significance.

4. If S is the set of points on the perimeter of a square, and C is the set of points on the circumference of the circumscribing circle, what meaning has $S \cap C$?

5. Is $\triangle$ a commutative operation? Justify your answer.

6. If $P = \{5, 8, 11, 14, 17, 19\}$, $Q = \{x \mid x = 2y + 3, y = 4, 5, 6, 7, 8\}$, $R = \{x \mid x = 3z - 1, z = 2, 3, 4, 5, 6\}$, find:
   (a) $P \cap (Q \triangle R)$ (b) $(P \cap Q) \triangle (P \cap R)$
   What general conclusion does the result suggest? Does it enable this conclusion to be drawn? Examine whether in this case symmetric difference distributes over intersection.

7. For the sets P, Q, R defined in question 6, find:
   (a) $P \cup (Q \triangle R)$ (c) $P \triangle (Q \cup R)$
   (b) $(P \cup Q) \triangle (P \cup R)$ (d) $(P \triangle Q) \cup (P \triangle R)$

Do the results suggest any possible generalization? Can a firm generalization be made?

8. Let $\mathscr{E}$ be the set of all people, C the set of all children, F the set of all females, M the set of all musical people.

  (i) What groups do the following sets represent:

  (a) C' (c) $F' \cap M$

  (b) $F - M$ (d) $F \cap C \cap M$

  (ii) Express in terms of C, F, M the groups:

  (a) of unmusical boys

  (b) of musical men

  (c) of musical children and unmusical adults

9. Draw a Venn diagram to verify the truth of each of the following statements:

  (a) If $P \subset Q$, then $P \cup Q = Q$

  (b) If $P \cap Q = \emptyset$, then $P \cup Q' = Q'$

  (c) $(P - Q) \subset (P \cup Q)$

  (d) If $P \subset Q$, then $P \cup (Q - P) = Q$

10. Let $P \subset S$ and $Q \subset S$. Show that:

  (a) if $P \subset (S - Q)$, then $P \cap Q = \emptyset$

  (b) if $(S - Q) \cap P = \emptyset$, then $P \subset Q$

11. Sets $\mathscr{E}$, A, B, C, D are represented by the shaded regions in Fig. 2.9.

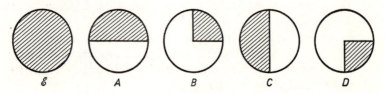

FIGURE 2.9

Show that:

(a) $A' - D = D' - A$ (c) $(A \,\Delta\, C) - B = A' - D$

(b) $(C \cup D) = B'$ (d) $(A \cap B) \cup C = D'$

12. Let $\mathscr{E}$ be the set of all positive integers, and let A, B be subsets of $\mathscr{E}$ where $A = \{n \,|\, 5 < n < 10\}$, $B = \{n \,|\, 2 < n < 8\}$.

  (i) Write out the sets:

           (a) $A' \cap B$ (b) $A \cap B'$ (c) $A' - B'$

  (ii) Write in set-builder notation the set $A' \,\Delta\, B'$

Exercises on Chapters 1 and 2

1. Write down eight simple identities involving either or both of the operators $\cap$ and $\cup$ and one or more of the sets $\emptyset$ (empty set), A, B, C, U (universal set) such that the identities remain true when $\cap$ and $\cup$ are replaced by $\times$ and $+$ and the sets $\emptyset$, A, B, C, U by the numbers 0, a, b, c, 1, respectively. (JMB)

2. Let P be the set of points on the circumference of a circle, I the set of points within the circle, L the set of points on a line, and let p_1, p_2, p_3, p_4 denote points. Draw a diagram to illustrate each of the following possibilities:

   (a) $L \cap I \cap P = \emptyset$ (c) $L \cap P = \{p_3\}$
   (b) $L \cap P = \{p_1, p_2\}$ (d) $L \cap P = \{p_1, p_2, p_3, p_4\}$

3. Verify that $(A - D) \subset [(A - B) \cup (B - C) \cup (C - D)]$

4. Write in roster notation the set:
   $A = \{x \mid 0 < x < 12,\ 2x^2 - 5x - 2 \text{ is a positive multiple of } 5\}$

5. Show by means of a counter example that symmetric difference is not distributive over relative complement.

6. Express the shaded regions in Fig. 2.10 (a)–(f) in terms of operations in A, B, C. Each set is represented by a rectangle, the enclosing rectangle being $\mathscr{E}$.

7. If P and Q are disjoint sets, and if $P \subset R$, $Q \subset R$, verify that $(R - P) - Q = R - (P \cup Q)$.

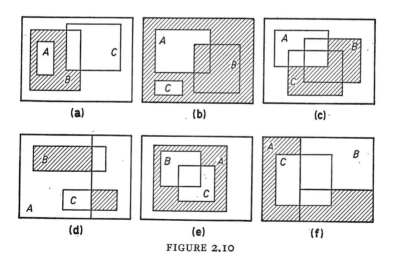

(a)   (b)   (c)

(d)   (e)   (f)

FIGURE 2.10

8. ABC is a triangle, S_{AB}, S_{BC}, S_{CA} are the sets of points in the sides AB, BC, CA. Interpret:

(a) $S_{AB} \cap S_{BC}$

(b) the set $S_P = \{P \mid P$ is a point equidistant from AB, $BC\}$; and the set $S_Q = \{Q \mid Q$ is a point equidistant from BC, $CA\}$

(c) $S_P \cap S_Q$

(d) $S_R \cap [S_{AB} \cup S_{BC} \cup S_{CA}]$, if $S_R = \{R \mid R$ is a point on the inscribed circle of $ABC\}$

9. P, Q, R, S are four sets such that $P \cup Q = R \cap S$.

(a) Draw a Venn diagram to illustrate the sets P, Q, R, S

(b) Show that $(R - P) \cap (S - Q) = \emptyset$

10. Refer to the note after question 7 in Exercise 2b (p. 16). Investigate how the four sets of values of a, b, c for which $a + bc = (a + b)(a + c)$ were obtained. Write down several more sets of such values.

3 | SOME SETS OF NUMBERS

Although the elements of a set may be objects of any kind, sets of numbers have an importance of their own: they enable answers to be given to certain questions which arise in situations common in everyday life.

3.1 The set of counting numbers (N)

The numbers with which we first become acquainted are those we use for counting, the set $N = \{1, 2, 3, 4, 5, \ldots\}$. In N there is an element which answers the question 'How many?' in relation to any finite collection of things. N is therefore known as the set of *counting numbers* (or, sometimes, *natural numbers*). Some, however, define the natural numbers differently, including zero in that set; so it is probably best to avoid the name, and instead to call N the set of counting numbers.

Let us suppose we have some coins. To find how many there are we count them, saying perhaps 'one, two, three, four, five, six'. But what is the meaning of this, and why does the recitation give us the answer to the question 'How many?'

Imagine we have a set of boxes the first of which contains one bean, the second two, the third three, and so on, and suppose the boxes are labelled with the symbols 1, 2, 3, . . . To find how many coins we have we may proceed as follows. We select any box and try to pair the beans it contains with the coins. If there are some beans for which there are no coins, or some coins for which there are no beans, we put the box back and try another.

When we find a box which has a bean for each of the coins, no more and no less, the number symbol on the label denotes the number of coins we have.

The process of counting therefore consists of establishing this kind of correspondence, called a *one-to-one correspondence*, between the objects to be counted and the elements of a standard set of objects (in this case beans in a box).

In Fig. 3.1 the downwards arrows indicate that for each bean there is a coin: the upward arrows show that for each coin there is a bean.

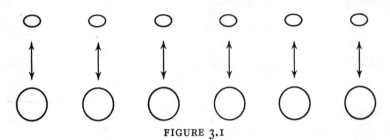

FIGURE 3.1

We can do without the boxes if we use the number symbols (or *numerals*) 1, 2, 3, . . . instead of beans. In the above example we shall then establish a one-to-one (1–1) correspondence between the coins and the elements of the set {1, 2, 3, 4, 5, 6}. In practice, of course, we memorize the number names 'one', 'two', 'three' and so on, and also the corresponding numerals 1, 2, 3 . . . Then when we repeat the number names from the beginning in order, one for each object being counted, the last number we name is the answer to the question 'How many?'

Order in N

In the last sentence, the phrases 'from the beginning' and 'in order' should be noted. We shall obtain the wrong answer when counting if we either fail to begin with 'one' or fail to name the the numbers in the proper order.

Consider the set $S = \{5, 3, 1, 8, 2, 6, 4, 7\}$, whose elements are arranged untidily; it is more natural to write $S = \{1, 2, 3, 4, 5, 6, 7, 8\}$ with the elements in ascending order of magnitude.

In the set N of counting numbers there is no element less than 1: 1 is the *leading element* of N. Also, if a, b are any two elements of N ($a \neq b$), there exist elements of N, m and n say, such that either $a = b + m$ or $a + n = b$. Thus, if $a = 23$, $b = 14$, m will be 9: if $a = 8$, $b = 20$, n will be 12.

If $a = b + m$, we say that a is *greater* than b: if $a + n = b$, we say that *a* is *less* than *b*. We are thus able to arrange the elements of any subset *S* of *N* in order of magnitude. If $a + 1 = b$, we call *b* the *successor* of *a*.

Example 1. Let α, β, γ, δ denote four counting numbers, and let $S = \{\alpha, \beta, \gamma, \delta\}$; then $S \subset N$. If the leading element of *S* is β, and $\alpha + \beta = \gamma$, $\beta + \delta = \alpha$, write the elements of *S* in ascending order of magnitude.

$$\alpha + \beta = \gamma \Rightarrow \alpha < \gamma \text{ and } \beta < \gamma$$
$$\beta + \delta = \alpha \Rightarrow \beta < \alpha \text{ and } \delta < \alpha$$

so either $\beta < \delta < \alpha < \gamma$ or $\delta < \beta < \alpha < \gamma$. But β is the leading element of *S*, so $S = \{\beta, \delta, \alpha, \gamma\}$.

Binary operations in N

Definition. If an operation $*$ on a pair of elements *a*, *b* of a set *S* produces the result $c \in S$, then $*$ is said to be a *binary operation* in *S*, and *S* is said to be *closed* under the operation $*$.

Example 2. Since the sum and product of any two counting numbers are counting numbers, *N* is closed under the operations of addition and multiplication. The operations denoted by $+$ and $\times$ are therefore binary operations in *N*. A counter example will show that subtraction is not a binary operation in *N*. For instance, if $5 - 8 = d$, then $d \notin N$; so *N* is not closed under subtraction.

Definition. If for all $x \in S$, $i \in S$ is such that $i*x = x*i = x$, then *i* is called the *identity element* in *S* for the operation $*$.

Example 3. Since for all $n \in N$, $1.n = n.1 = n$, 1 is the identity element in *N* for multiplication.

Exercise 3a

1. Provide a counter example to show that *N* is not closed under division.
2. Is there an identity element in *N* for addition? (Ask: if $a \in N$, $b \in N$, and $a + b = b$, what, if any, is the value of *a*?)

B

3. Let $A \subset N$, and let the elements of A be a, b, c, d, e, f. If $b + c = f$, $a + c = d$, $b + f = a$, $b + e = d$, $c + f = e$, $c + e = a$, $c + c = b$, write A with its elements in ascending order of magnitude.
4. Is the operation of finding the average of two numbers a binary operation in N?
5. If $a \in N$, $b \in N$, is the operation $*$ defined by $a*b = a^2 + ab + b^2$ a binary operation in N?

Finite induction

Suppose we have a set S of counting numbers. Under what conditions will S contain all the counting numbers? That is, how can we determine whether or not $S = N$ is a true statement?

The principle of finite induction gives us an answer to this question. It states that if S is a set of counting numbers in which:

(*i*) $1 \in S$

(*ii*) $k \in S \Rightarrow (k + 1) \in S$

then $S = N$.

The truth of this is evident since, from (*ii*), $1 \in S \Rightarrow 2 \in S$, and $2 \in S \Rightarrow 3 \in S$, and so on. Thus S contains all the counting numbers and no others, and hence $S = N$.

This principle provides an interesting method of proof, applicable to some types of problem relating to counting numbers, of which the following are examples.

Example 4. Prove that $1^2 + 2^2 + 3^2 + \ldots + n^2 = \frac{1}{6}n(n + 1)(2n + 1)$ for all $n \in N$.

Let S be the set of all values of n for which the statement is true. If we can show that S satisfies conditions (*i*) and (*ii*) above, we can say $S = N$, and the proof will be complete.

(*i*) $1 \in S$ because when $n = 1$, $\frac{1}{6}n(n + 1)(2n + 1) = 1 = 1^2$.

(*ii*) If $k \in S$, then $1^2 + 2^2 + 3^2 + \ldots + k^2 = \frac{1}{6}k(k + 1)(2k + 1)$.

So, adding $(k + 1)^2$ to both sides:

$$1^2 + 2^2 + \ldots + k^2 + (k + 1)^2 = \frac{1}{6}k((k + 1)(2k + 1) + (k + 1)^2$$
$$= \frac{1}{6}(k + 1)(2k^2 + 7k + 6)$$
$$= \frac{1}{6}(k + 1)(k + 2)(2k + 3)$$
$$= \frac{1}{6}(k + 1)[(k + 1) + 1][2(k + 1) + 1]$$

which is of the given form with $(k + 1)$ for k.

Thus, if $k \in S$, then $(k + 1) \in S$.

Since S satisfies both conditions, $S = N$ and the result is true for all $n \in N$.

Example 5. Prove that, for all $n \in N$, $2^{3n} - 1$ is divisible by 7.

Let S be the set of all counting numbers n for which $2^{3n} - 1$ is divisible by 7. Then:

(*i*) $1 \in S$, for, when $n = 1$, $2^{3n} - 1 = 7$.

(*ii*) When $n = (k + 1)$:
$$2^{3n} - 1 = 2^{3k+3} - 1 = 8 . 2^{3k} - 1 = (2^{3k} - 1) + 7 . 2^{3k}$$
This is divisible by 7 if $(2^{3k} - 1)$ is divisible by 7: thus
$$k \in S \Rightarrow (k + 1) \in S.$$

Since S satisfies both conditions, $S = N$ and $2^{3n} - 1$ is divisible by 7 for all $n \in N$.

The conclusiveness of this argument should be noted. It would be possible to show that the result holds for any given value of n: *e.g.*, in Example 5, when $n = 4$, $2^{3n} - 1 = 2^{12} - 1 = 4095$, which is 7×585. But establishing the truth of the statement for specific values of n, no matter how many, does not prove it true for all counting numbers n (*cf.* the Note to question 7 in Exercise 2b on p. 16). The second condition of the principle of finite induction, if satisfied, means that any element of S also has its successor in S, from which it follows that, provided $1 \in S$, all elements of N are in S.

Exercise 3b

Use the principle of finite induction to prove the truth of each of the following statements. In each case, $n \in N$.

1. $1 + 2 + 3 + \ldots + n = \frac{1}{2}n(n + 1)$
2. $5.6 + 6.7 + 7.8 + \ldots$ to n terms $= \frac{1}{3}n(n^2 + 15n + 74)$
3. $n^3 - n$ is divisible by 3
4. $\frac{1}{2}n(n + 1)(2n + 1)$ is divisible by 3, by showing that:
$$\tfrac{1}{2}(n + 1)(n + 2)(2n + 3) = \tfrac{1}{2}n(n + 1)(2n + 1) + 3(n + 1)^2$$

3.2 The set of integers (Z)

The counting numbers provide solution sets for many problems, but not by any means for all. Study the following table: in each case $x \in N$.

	Equation	Solution set
1.	$4x - 1 = 19$	$\{5\}$
2.	$x^2 - 11x + 28 = 0$	$\{4, 7\}$
3.	$x + 7 = 3$	$\emptyset$
4.	$2(x + 4) = x + 8$	$\emptyset$
5.	$2x^2 - 5x - 3 = 0$	$\{3\}$

To solve equation **3**, we should have to subtract 7 from 3 which is meaningless if $x \in N$. In equation **4**, $x = 0$, but $0 \notin N$. In equation **5**, $x = 3$ or $-\frac{1}{2}$, but only the solution 3 belongs to N.

It therefore becomes necessary to extend our number system by introducing new numbers, one of them being a number for $3 - 7$. This may be done by using the set N in the following way.

Let us write the number $3 - 7$ in the form $(3, 7)$. This is called an *ordered pair*: the order is important, since $(7, 3)$ would represent $7 - 3$ (*i.e.*, 4).

We will say that the two numbers represented by the ordered pairs (a, b) and (c, d) are *equal* if $a + d = b + c$. Thus, $(3, 7) = (4, 8) = (16, 20) = \ldots = (k, k + 4)$ for all $k \in N$. There is thus a limitless number of ways in which the number $3 - 7$ may be written as an ordered pair. This is just another way of saying that there is an infinite number of equations all having the same solution set, *e.g.*, $x + 7 = 3$, $x + 8 = 4$, $\ldots$, $x + 20 = 16$, and so on.

We call this extended number system, including the counting numbers and our new numbers, the set of *integers*, and denote this set by Z.

Exercise 3c

1. Write in several simpler forms the integers represented by:

   (*a*) $(52, 18)$ (*b*) $(17, 71)$ (*c*) $(32, 32)$

2. If $(5, 9)$ and $(3, 3 + k)$ are equal integers, what is the value of k?
3. Show that the integers represented by $(p, q + 1)$ and $(p + 2, q)$ cannot be equal.
4. Write as ordered pairs the solutions of the equations:

   (*a*) $x + 11 = 5$ (*c*) $8 = 15 + x$

   (*b*) $61 + x = 72$

In each case express the solution as the simplest possible ordered pair. [Remember that in (a, b), $a \in N$ and $b \in N$.]

Addition and multiplication of integers

We define the sum and product of two integers (a, b) and (c, d) thus:

$$(a, b) + (c, d) = (a + c, b + d)$$
$$(a, b) \cdot (c, d) = (ac + bd, ad + bc)$$

It will be seen in the following paragraphs that these definitions are not arbitrary, but have been chosen with a definite end in view.

Example 6

(i) $(3, 5) + (1, 6) = (4, 11) = (1, 8)$. Why?

(ii) $(3, 5) \cdot (1, 6) = (33, 23) = (11, 1)$. Why?

Exercise 3d

Write in its simplest form as an ordered pair the result of each of the following operations in integers.

1. $(2, 5) + (17, 11) + (8, 9)$
2. $(2, 3) \cdot (3, 4)$
3. $(4, 2) \cdot [(3, 7) + (6, 3)]$
4. $(4, 2) \cdot (3, 7) + (4, 2) \cdot (6, 3)$

What conclusion do you draw from the last two results? Is it a firm conclusion, or is it only possible?

The positive integers (Z^+)

In the integer (a, b) we may have $a > b$, $a = b$, or $a < b$. If $a > b$, we may write $a = b + m$ $(m \in N)$, so that $(a, b) = (b + m, b)$.

From the definition of equality, $(b + m, b) = (c + m, c)$, which means that, for any given value of $m \in N$, $(b + m, b)$ has a value independent of the value of b. We may therefore establish a 1–1 correspondence between ordered pairs (a, b), in which $a > b$, and the counting numbers, thus:

$(b + 1, b)$	$(b + 2, b)$	$(b + 3, b)$	...	$(b + m, b)$	...
$\updownarrow$	$\updownarrow$	$\updownarrow$	...	$\updownarrow$	...
1	2	3	...	m	...

We may go further. If we add two such ordered pairs, say $(b + x, b)$ and $(b + y, b)$, we have the ordered pair $(2b + x + y, 2b)$ which is equal to $(b +x +y,\ b)$ by the definition of equality. So we have the 1–1 correspondence:

$$(b +x,\ b) +(b +y,\ b) = (b +x +y,\ b)$$
$$\updownarrow \qquad\qquad \updownarrow \qquad\qquad\quad \updownarrow$$
$$x \quad + \quad y \quad = \quad x +y \qquad\qquad (3.1)$$

The same is true of multiplication, for:
$$(b +x,\ b).(b +y,\ b) = (2b^2 +bx +by +xy,\ 2b^2 +bx +by)$$
$$= (b +xy,\ b) \quad \text{by the definition of equality}$$
So we have the 1–1 correspondence:

$$(b +x,\ b).(b +y,\ b) = (b +xy,\ b)$$
$$\updownarrow \qquad\quad \updownarrow \qquad\quad \updownarrow$$
$$x \quad . \quad y \quad = \quad xy \qquad\qquad (3.2)$$

Equations (3.1) and (3.2) show that numbers of the form $(a,\ b)$ where $a > b$ behave exactly as counting numbers, and they may therefore be regarded as different symbols for the elements of the set N. For this reason we often refer to the counting numbers as the *positive integers*. The symbol Z^+ for the positive integers is self-explanatory. Z^+ and N are two symbols for the same set.

An isomorphism

In showing that numbers of the form $(b +m,\ b)$ where $b \in N$, $m \in N$ are in 1–1 correspondence with the positive integers and that the operations of addition and multiplication defined in Z^+ are preserved in the correspondence, we have established an *isomorphism* between the numbers $(b +m,\ b)$ and the elements of Z^+.

Definition. Two mathematical systems A, B are *isomorphic* if:
  (*i*) there is a 1–1 correspondence between the elements of the sets A, B. This is written $A \leftrightarrow B$;
  (*ii*) any operations and relations which are defined in A and B are preserved in the correspondence.

Both conditions must be satisfied. Thus, for example, there is a 1–1 correspondence between the elements of N and the even positive integers:

$$
\begin{array}{ccccccc}
1 & 2 & 3 & 4 & 5 & 6 \ldots n \ldots \\
\updownarrow & \updownarrow & \updownarrow & \updownarrow & \updownarrow & \updownarrow \qquad\quad \updownarrow \\
2 & 4 & 6 & 8 & 10 & 12 \ldots 2n \ldots
\end{array}
$$

Also
$$1 + 3 = 4$$
$$\updownarrow \quad \updownarrow \quad \updownarrow$$
$$2 + 6 = 8$$

so that the correspondence is preserved in addition. But it is not preserved in multiplication:
$$1 \times 3 = 3$$
$$\updownarrow \quad \updownarrow$$
$$2 \times 6 = 12$$

Here 3 does not correspond to 12. Hence condition (ii) is not completely satisfied, and the two systems are not isomorphic.

Exercise 3e

1. Examine whether the set N is isomorphic to the set of odd positive integers.
2. 'The set N is isomorphic to the set $A = \{x \mid x = y + 2, y \in N\}$.' Is this statement true or false?
3. Is the set $\{un, deux, trois, quatre, cinq, six, sept, \ldots\}$ isomorphic to Z^+?
4. Prove that $(5, 2).(8, 5) = (50, 41)$ and set up a 1–1 correspondence with an operation in Z^+.
5. If $(b + x, b).(b + y, b) = (25, 1)$, tabulate the possible values of the positive integers x, y.

Zero

Consider now the case in which $a = b$ in the pair (a, b). We may then write the pair as (a, a).

Two results follow:

(i) $(a, a) + (c, d) = (a + c, a + d) = (c, d)$ by the definitions of addition and equality

(ii) $(a, a).(c, d) = (ac + ad, ad + ac) = (a, a)$ by the definitions of multiplication and equality

(i) shows that (a, a) is an identity element for addition for all $n \in N$. It is therefore a number not in N (question 2 in Exercise 3a, p. 27). We call it *zero*, and denote it by 0.

(ii) shows that $0.m = 0$ where $m \in Z$.

The negative integers (Z^-)

Now suppose that in the ordered pair (a, b) $a < b$: we may then write $b = a + n$ and $(a, b) = (a, a + n)$.

If we add $(a, a + n)$ and $(a + n, a)$ we have $(2a + n, 2a + n)$ which is zero. (Why?). But $(a + n, a)$ represents the positive integer n: so $(a, a + n)$ is a number that, added to $(a + n, a)$, gives zero. It is called the *additive inverse* of $(a + n, a)$ or n.

We usually write the additive inverse of n as $-n$.

This suggests that ordered pairs of the form (a, b) in which $a < b$ may be regarded as different symbols for the *negative integers*; but it has yet to be shown that multiplication is preserved in the correspondence.

Consider $(a, a + n).(b, b + p)$. The product is:

$$[ab + (a + n)(b + p), a(b + p) + b(a + n)]$$
$$= [(2ab + ap + bn) + np, (2ab + ap + bn)]$$
$$= (k + np, k) \text{ where } k = 2ab + ap + bn$$

So we have the correspondence preserved:

$$(a, a + n).(b, b + p) = (k + np, k)$$
$$\updownarrow \qquad\quad \updownarrow \qquad\qquad \updownarrow$$
$$-n \quad . \quad -p \quad = \quad np$$

There is thus an isomorphism between the set of ordered pairs (a, b), where $a < b$, and the negative integers.

Exercise 3f

1. Write down the integers represented by the ordered pairs:
   (a) $(14, 8)$ (d) $(q, q + 1)$
   (b) $(6, 11)$ (e) $(17, 17)$
   (c) $(p + 6, p)$ (f) $(a + b + c, a + b + c + 9)$
2. Write as integers the results of the following operations:
   (a) $(3, 2) + (2, 5) + (1, 6)$
   (b) $(7, 3).(2, 5)$
   (c) $(5, 8).(5, 8) + (7, 4).(4, 7)$
3. By considering the product $(a, a + n)(b + m, b)$, show that the product of a negative integer and a positive integer is a negative integer.
4. Show that the product of two integers (a, b) and (c, d) (p. 31) agrees with the result of multiplying $(a - b)$ by $(c - d)$.

We have seen that, by means of ordered pairs of counting numbers, we may extend our number system to include zero and the negative integers. We shall now revert to the familiar notation $-k$ for negative integers $(k \in N)$ rather than continuing to use the ordered pair $(a, a + k)$.

Subtraction of integers

The set N of positive integers is closed under two operations, addition and multiplication, but not under subtraction or division. The set Z of integers is also closed under addition and multiplication, and also under subtraction, since $a \in Z$ and $b \in Z \Rightarrow (a - b) \in Z$. Subtraction is a binary operation in integers.

If $a \in Z$, $b \in Z$ and $a - b$ is positive, we say $a > b$; if $a - b$ is negative, $a < b$; and if $a - b$ is zero, $a = b$. The elements of Z can thus be ordered:

$$Z = \{\ldots -4, -3, -2, -1, 0, 1, 2, 3, 4, \ldots\}$$

Z may be represented by points in line equally spaced:

← .   .   .   . →
  -5   -4   -3   -2   -1   0   1   2   3   4   5

Note that Z has no leading element.

Exercise 3g

1. Justify each of the following statements:
   (a) $-8 < 1$ (c) $0 < n$ $(n \in N)$
   (b) $-2 > -20$ (d) $0 > -5$
2. If $a < 0$, $b > 0$, prove that $a - b < 0$.
3. If $x - y - 2z < 0$, $y < 0$ prove that $(xy + z^2) > (y + z)^2$.
4. Is division a binary operation in integers? Justify your answer.

Division of integers

It will have been found that division is not a binary operation in the set Z. This may be proved by a counter example: if $17 \div 3 = u$, then $u \notin Z$.

An integer p is divisible by an integer q if there exists an integer r such that $p = qr$. We then say that p is a multiple of q and q is a *divisor* of p. This is written $q \mid p$ (read as 'q divides p').

Example 7. The statements $5 \mid 30$, $-4 \mid 28$, $3 \mid -18$, $-7 \mid -7$, $6 \mid 0$, $0 \mid 0$ are all true. The statements $15 \mid -3$, $-8 \mid 20$, $0 \mid -5$ are all untrue.

If $p = qr$ and $q \neq 0$, the integer r is uniquely determined by the relation $q \mid p$. If $p = q = 0$, then, while $q \mid p$ is a true statement, it does not define an unique integer r since $0 . x = 0$ for all x.

An integer i is *even* if $i = 2a$ $(a \in Z)$: it is *odd* if $i = 2a - 1(a \in Z)$.

If i_1, i_2 are two even integers, say $i_1 = 2a$, $i_2 = 2b$, then $i_1 i_2$ is an even integer $(i_1 i_2 = 4ab)$. Conversely, if $i_1 i_2$ is even ($= 2pq$ say), then i_1 or i_2 (or both) will be even.

Exercise 3h

Make and justify statements about $i_1 i_2$:
(*a*) where i_1, i_2 are both odd   (*b*) where i_1 is odd and i_2 even

A non-zero integer p is *prime* if:
(*i*) $p \neq \pm 1$
(*ii*) p has no divisors other than ± 1 and $\pm p$

Thus, the smallest positive prime p_1 is 2: it is followed by $p_2 = 3$, $p_3 = 5$, $p_4 = 7$, $p_5 = 11$, and so on. Write down p_6, p_7, p_8, $\ldots p_{20}$. It is not possible by studying these twenty primes to say what the value of p_{21} would be. Indeed, it might appear possible that there is a largest prime, and that we could have no guarantee that p_{21} exists. It is easy to prove that it does exist, and that, in fact, there is no largest prime.

Theorem I

The number of positive primes is infinite.

This may be proved by an indirect method. We assume the conclusion to be false, and show that this assumption leads to a contradiction.

Proof. Suppose there are exactly n primes p_1, p_2, p_3, $\ldots p_n$. Consider the number $P = (p_1 . p_2 . p_3 . \ldots p_n) + 1$. P is greater than each of the n primes p_1, p_2, p_3, $\ldots p_n$ and is not divisible by any one of them: hence it is not divisible by any integer other than itself and 1. So by definition P is a prime number, and $P > p_n$ which is contrary to the supposition that p_n is the greatest prime. Hence there is an infinity of positive primes.

3.3   The set of rationals (Q)

Even with the extension of our number system to include all integers, there are still problems for which we have no solution. Thus, the solution set of the equation $2x - 1 = 6$ $(x \in Z)$ is $\emptyset$, because 2 is not a divisor of 7.

To enable such an equation to be solved, we must make a further extension beyond Z to the set Q of *rational numbers*. A rational number is the *ratio* of two integers a, b ($b \neq 0$). Thus $\frac{2}{3}$, $\frac{-1}{10}$, $\frac{5}{-1}$ are rationals. 0.67 can be written as a ratio, $\frac{67}{100}$, and is therefore a rational; so also is the recurring decimal $0.\dot{3}$, being equal to $\frac{1}{3}$.

We extended our number system from counting numbers to integers by means of an ordered pair of counting numbers. We are again making an extension by using an ordered pair of numbers, this time, of integers. We could write the rational $\frac{a}{b}$ as (a, b): but as the former notation is already familiar it is simpler to adopt it. For every rational $\frac{a}{b}$ it is understood that $b \neq 0$, since $\frac{a}{0}$ has no numerical value.

Equality of rationals

Two rationals $\frac{a}{b}$, $\frac{c}{d}$ are equal if, and only if, $ad = bc$. This definition will be familiar from experience with arithmetical fractions.

There are thus any number of forms in which a given rational may be written: *e.g.*, $\frac{3}{4} = \frac{9}{12} = \frac{21}{28} = \ldots$ Of these rationals, $\frac{3}{4}$ is in its *lowest terms*: all the rationals equal to $\frac{3}{4}$ are *reducible* to $\frac{3}{4}$. This is just another way of saying that many equations may have the same solution set, *e.g.*, $4x = 3$, $12x = 8$, $28x = 21$, $\ldots$

If $\frac{a}{b}$, $\frac{c}{d}$ are two rationals, and $\frac{a}{b} - \frac{c}{d}$ is positive, then $\frac{a}{b} > \frac{c}{d}$, *i.e.*, $ad > bc$. If $\frac{a}{b} - \frac{c}{d}$ is negative, $\frac{a}{b} < \frac{c}{d}$, *i.e.*, $ad < bc$.

Sum, difference and product of rationals

The definitions of addition, subtraction and multiplication will be familiar:

(*i*) $\dfrac{a}{b} + \dfrac{c}{d} = \dfrac{ad + bc}{bd}$

(*ii*) $\dfrac{a}{b} - \dfrac{c}{d} = \dfrac{ad - bc}{bd}$

(*iii*) $\dfrac{a}{b} \cdot \dfrac{c}{d} = \dfrac{ac}{bd}$

From (*i*) an identity element for addition is $\dfrac{0}{1}$ since, for all $a, b \in Z$ except $b = 0$, $\dfrac{a}{b} + \dfrac{0}{1} = \dfrac{a}{b}$.

Also, since $\dfrac{a}{b} + \dfrac{-a}{b} = \dfrac{0}{1}$, $\dfrac{-a}{b}$ is the additive inverse of $\dfrac{a}{b}$.

From (*iii*) $\dfrac{1}{1}$ is an identity element for multiplication since, for all $a, b \in Z$ except $b = 0$, $\dfrac{a}{b} \cdot \dfrac{1}{1} = \dfrac{a}{b}$.

Since $\dfrac{0}{1} = \dfrac{0}{2} = \dfrac{0}{3} = \ldots = \dfrac{0}{k}$ and $\dfrac{1}{1} = \dfrac{2}{2} = \dfrac{3}{3} = \ldots = \dfrac{k}{k}$ $(k \in N)$, the additive and multiplicative identity elements are expressible in an infinity of forms.

The set Q is closed under the operations of addition, subtraction and multiplication. This is clear from definitions (*i*) to (*iii*) above, since each of these three operations on two rationals gives a rational.

Division of rationals

A rational $\dfrac{a}{b}$ is divisible by a non-zero rational $\dfrac{c}{d}$ if there exists a rational $\dfrac{x}{y}$ such that $\dfrac{a}{b} = \dfrac{c}{d} \cdot \dfrac{x}{y}$: that is, if $\dfrac{x}{y} = \dfrac{ad}{bc}$.

Now $b \neq 0$, and, since $\dfrac{c}{d} \neq 0$, $c \neq 0$, so $bc \neq 0$: hence $\dfrac{ad}{bc}$ exists. Thus any rational number is divisible by any non-zero rational.

It follows that the set of non-zero rationals is closed under the four operations denoted by $+$, $-$, $\times$, $\div$.

Example 8. Establish an isomorphism between Z and rationals of the form $\dfrac{a}{1}$ $(a \in Z)$.

We have to show that two conditions are satisfied (p. 32).

(*i*) There is clearly a 1–1 correspondence between the two sets:

$$\ldots \dfrac{-2}{1}, \dfrac{-1}{1}, \dfrac{0}{1}, \dfrac{1}{1}, \dfrac{2}{1}, \dfrac{3}{1}, \ldots \dfrac{k}{1}, \ldots$$
$$\updownarrow \quad \updownarrow \quad \updownarrow \quad \updownarrow \quad \updownarrow \quad \updownarrow \qquad \updownarrow$$
$$\ldots -2, \ -1, \ 0, \ 1, \ 2, \ 3, \ \ldots k, \ldots$$

(*ii*) There are three binary operations in Z: addition, subtraction and multiplication. These are all preserved in the correspondence, thus:

$$\frac{a}{1} + \frac{b}{1} = \frac{a+b}{1}$$
$$\updownarrow \quad \updownarrow \qquad \updownarrow$$
$$a + b = a + b$$

$$\frac{a}{1} - \frac{b}{1} = \frac{a-b}{1}$$
$$\updownarrow \quad \updownarrow \qquad \updownarrow$$
$$a - b = a - b$$

$$\frac{a}{1} \cdot \frac{b}{1} = \frac{ab}{1}$$
$$\updownarrow \quad \updownarrow \qquad \updownarrow$$
$$a.b = ab$$

Hence the isomorphism is established.

Note. It is not really necessary to test subtraction, since the existence of additive inverses in the integers and the rationals enables a difference to be expressed as a sum: thus, $\frac{a}{1} - \frac{b}{1}$ is the same as $\frac{a}{1} + \frac{-b}{1}$; and $a - b$ as $a + (-b)$.

Example 9. Show that between any two unequal rationals $\frac{a}{b}, \frac{c}{d}$ there is an infinity of rationals.

If $\frac{a}{b} \neq \frac{c}{d}$, one of them is the greater. Let $\frac{c}{d} > \frac{a}{b}$: then $bc > ad$. Consider the rational $\frac{a+c}{b+d} - \frac{a}{b}$. It is $\frac{bc-ad}{b(b+d)}$, which is positive since $\frac{c}{d} > \frac{a}{b}$: hence $\frac{a+c}{b+d} > \frac{a}{b}$.

Also, $\frac{c}{d} - \frac{a+c}{b+d} = \frac{bc-ad}{d(b+d)}$, which is positive, since $\frac{c}{d} > \frac{a}{b}$: hence $\frac{c}{d} > \frac{a+c}{b+d}$. So $\frac{a}{b} < \frac{a+c}{b+d} < \frac{c}{d}$.

By the same argument we may prove that $\dfrac{a}{b} < \dfrac{2a+c}{2b+d} < \dfrac{a+c}{b+d}$, and this may be continued indefinitely. Hence there is an infinity of rationals between $\dfrac{a}{b}$ and $\dfrac{c}{d}$.

Exercise 3i

1. Write down ten rationals which are between $\frac{9}{10}$ and $\frac{10}{11}$ in value.
2. Give an example to illustrate the truth of the statement, 'In Q, multiplication distributes over addition.'
3. If $a, b, c, d \in N$, $b > a$, $c > d$, prove that $\dfrac{a}{b} < \dfrac{ac}{bd} < \dfrac{c}{d}$.
   What is the corresponding relation when $b < a$, $c < d$? And when $a = b$, $c = d$?
4. Show that there is a 1–1 correspondence between the set A of rationals $\dfrac{1}{1}, \dfrac{1}{2}, \dfrac{1}{3}, \ldots \dfrac{1}{k}, \ldots$ and the set N, but that A is not isomorphic to N.

3.4   The set of real numbers (R)

We have proved that between any two unequal rationals $\dfrac{a}{b}, \dfrac{c}{d}$ no matter by how little they may differ in value, there is an infinity of rationals. It might seem, therefore, that if the rationals are represented by points in line, then every point on the line would represent a rational. For example, $\frac{204}{317}$ and $\frac{205}{317}$ differ by only $\frac{1}{317}$; but an infinite number of rationals lie between them in value.

However, the rationals do *not* complete the number line: for, while every rational is represented by a point on the line, not every point on the line represents a rational.

Example 10. Show:
   (*i*) that $\sqrt{2}$ (the positive square root of 2) is represented by a point on the number line
   (*ii*) $\sqrt{2}$ is not a rational

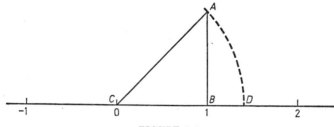

FIGURE 3.2

(*i*) If in Fig. 3.2 $AB = 1$, then $CA = \sqrt{2}$: so, if $CD = CA$, D on the number line represents $\sqrt{2}$.

(*ii*) may be proved indirectly. Suppose $\dfrac{\sqrt{2}}{1} = \dfrac{a}{b}$, where $\dfrac{a}{b}$ is a rational in its lowest terms. Then:

$\dfrac{2}{1} = \dfrac{a^2}{b^2} \Rightarrow a^2 = 2b^2$ (definition of equality)

$\qquad \Rightarrow a^2$ is an even integer (definition on p. 36)
$\qquad \Rightarrow a$ is an even integer
$\qquad \Rightarrow a = 2m \; (m \in Z)$
$\qquad \Rightarrow a^2 = 4m^2$
$\qquad \Rightarrow b^2 = 2m^2$
$\qquad \Rightarrow b^2$ is an even integer
$\qquad \Rightarrow b$ is an even integer

$\qquad \Rightarrow \dfrac{a}{b}$ is reducible

which contradicts the supposition that $\dfrac{a}{b}$ is in its lowest terms.

Hence $\sqrt{2}$ is an *irrational* number: it cannot be written as the ratio of two integers.

Other examples of irrationals are π, log 6, sin $\frac{1}{3}\pi$. Rational values of these are only approximations (*e.g.*, $\frac{22}{7}$ for π).

The set of rationals together with the set of irrationals make up the set R of *real numbers*. R is closed under all four operations $+$, $-$, $\times$, $\div$ (with the necessary exception of division by zero).

The question may arise whether there may still be points on the number line which do not represent either rationals or irrationals. It can be shown that this is not the case, and we may speak of the number line as the *real number line*. Every point of the line represents a real number: every real number is represented by a point of the line.

With the introduction of irrationals the range of our ability to solve problems has been extended still further. Thus, if $5x^2 = 11$ ($x \in R$), we have the solution set $\{+\sqrt{2 \cdot 2}, -\sqrt{2 \cdot 2}\}$.

The following table illustrates how the progressive extension of our number system from the counting numbers N to the integers Z, then to the rationals Q, and now to the real numbers R, has increased our ability to solve problems.

	Equation		Solution set		
		if $x \in N$	if $x \in Z$	if $x \in Q$	if $x \in R$
1.	$x + 8 = 3$	$\emptyset$	$\{-5\}$	$\{-5\}$	$\{-5\}$
2.	$2x + 8 = 3$	$\emptyset$	$\emptyset$	$\{-2 \cdot 5\}$	$\{-2 \cdot 5\}$
3.	$2x^2 - 8 = 3$	$\emptyset$	$\emptyset$	$\emptyset$	$\{\sqrt{5 \cdot 5}, -\sqrt{5 \cdot 5}\}$
4.	$2x^2 + 8 = 3$	$\emptyset$	$\emptyset$	$\emptyset$	$\emptyset$

To solve equation **4** we should need still another extension of our number systems to *complex numbers* which are dealt with separately in the book of that title in this series.

Exercises on Chapter 3

1. Let $a, b \in Z$, and let an operation $*$ in Z be defined by the equation $a*b = a^3 - b^3$. Is $*$ a binary operation in Z?
2. If $a, b \in Q$ and $a*b = \sqrt{ab}$, is $*$ a binary operation in Q? Is $*$ a binary operation in R?
3. Prove by finite induction that, for all $n \in N$:
   $1.3 + 2.4 + 3.5 + \ldots$ to n terms $= \frac{1}{6}n(n+1)(2n+7)$
4. Prove by finite induction that, for all $n \in N$:
   $$(1^3 - 1) + (2^3 - 2) + (3^3 - 3) + \ldots + (n^3 - n)$$
   $$= \frac{1}{4}n(n+1)(n+2)(n-1)$$
5. Prove that, if n is any positive integer, $3^{2n} + 7$ is divisible by 8.
6. Write out the solution sets of the following equations:
   (a) $25x^2 - 1 = 8$ $(x \in R)$ (d) $(x + \pi)(2x - \pi) = 0$ $(x \in Q)$
   (b) $(3x - 5)(x + 7) = 0$ $(x \in Z)$ (e) $x(x - 1)(2x + 1) = 0$ $(x \in N)$
   (c) $\dfrac{5}{x} = \dfrac{7}{x-1}$ $(x \in Z)$ (f) $2x^2 - 3x + 1 = 0$ $(x \in Z)$
7. Draw a diagram showing a 1–1 correspondence between N and

the set $A = \{x \mid x = y^2, y \in N\}$. Are addition and multiplication preserved in the correspondence? Is N isomorphic to A?

8. If $a \mid b$ and $c \mid d$, prove that $ac \mid bd$. Is $(a + c) \mid (b + d)$ also true?

9. Prove each of the following statements false:
   (a) The sum of two prime numbers is a prime number.
   (b) If p is a prime, then $2p^2 + 1$ is a prime.
   (c) If two numbers are relatively prime (*i.e.*, have no common divisor other than 1), then their sum and difference are relatively prime.

10. If $\dfrac{a}{b} < \dfrac{c}{d}$, prove $\dfrac{a}{b} < \sqrt{\dfrac{ac}{bd}} < \dfrac{c}{d}$.

Exercises on Chapters 1–3

1. If $M_2 = \{x \mid x = 2y, \ y \in N\}$, $M_3 = \{x \mid x = 3y, \ y \in N\}$, define the set $M_2 \Delta M_3$ in set-builder notation. Write down the nine elements of lowest value in $M_2 \Delta M_3$. Given that $(6n - 2) \in (M_2 \Delta M_3)$, what are the next lower- and next higher-valued elements?

2. Prove that, for all $n \in N$, $1^3 + 2^3 + 3^3 + \ldots + n^3 = \frac{1}{4}n^2(n + 1)^2$.

3. Let θ and ϕ be two operations defined in N as follows:
   $$a \ \theta \ b = 2a + b, \qquad a \ \phi \ b = 2ab$$
   (a) Are θ and ϕ commutative operations?
   (b) Are θ and ϕ associative operations?
   (c) Prove $a \ \phi \ (b \ \theta \ c) = (a \ \phi \ b) \ \theta \ (a \ \phi \ c)$, and state this result in words.
   (d) Examine whether θ is distributive over ϕ.

4. In a game of Twenty Questions the team were asked to guess a number. The first four questions they asked were, in order:
   (a) Is it an even number?   (c) Is it a rational number?
   (b) Is it positive?   (d) Is it $-e$?
   To each of these questions the answer was 'No'. Do you consider that the team made a skilful start? If not, suggest a better set of four opening questions. How does set theory assist in this?

5. Prove that it would be fruitless to try to prove that $p^6 - 1$ is divisible by 168 when p is any prime number greater than 3.

6. Prove carefully that $\sqrt{3} \notin Q$.

7. Find and tabulate all the solutions of the equation $5x + 4y = 68$, where $x, y \in N$.

8. In Fig. 3.3, *AB* and *CD* are any two line segments. Prove that the points of *AB* can be put in a 1–1 correspondence with the points of *CD*. Deduce that any two line segments have the same number of points.

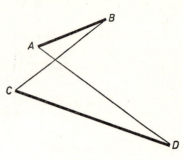

FIGURE 3.3

9. Prove that in the set of integers the operation 'to the power of' is not associative: *i.e.*, that $(x^y)^z \neq (x)^{y^z}$.

10. Prove that:

   (*a*) there are infinitely many rational numbers *r* such that $1 < r < 2$

   (*b*) there is no rational number $\dfrac{p}{q}$ such that $\left(\dfrac{p}{q}\right)^2 = 5$ (JMB)

11. If *a*, *b* are rational numbers such that $0 < a < b$, prove that:

   (*a*) $0 < \dfrac{1}{b} < \dfrac{1}{a}$

   (*b*) there exists a rational number *c* such that $a < c < b$

   (*c*) for any rational number *d*, $a + d < b + d$ (JMB)

12. Let *X* be the set of all numbers of the form $m + n\sqrt{2}$, where *m* and *n* are integers greater than zero. Determine whether *X* is closed under each of the operations subtraction, multiplication and division.

   Find also whether *X* is closed under the operation of taking the positive square root.   (JMB)

4 | PRODUCT SETS

4.1 Ordered pairs

We have already seen how an ordered pair (a, b) of positive integers can be used to introduce new numbers, zero and the negative integers as well as rationals. While this use of ordered pairs may have appeared strange, the idea of a pair is doubtless familiar from previous experience of graph-drawing. We speak of 'the point $(2\cdot5, 3)$', that is, the point at which $x = 2\cdot5$, $y = 3$, and we plot it in position P (Fig. 4.1).

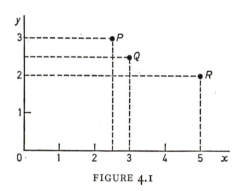

FIGURE 4.1

It is clear that $(2\cdot5, 3) \neq (3, 2\cdot5)$, for P and Q are not the same point. The pair is *ordered*.

Again, each ordered pair (a, b) is represented by *one* point in the plane, and each point in the plane represents *one* ordered pair. If R is the point (a, b), then $a = 5$, $b = 2$, and these are the only values of a, b which give a pair represented by R.

In general, if (a, b) and (c, d) are two ordered pairs, $(a, b, c, d \in R)$, then $(a, b) = (c, d)$ if and only if $a = c$ and $b = d$.

Note the distinction between the pair (a, b) and the set $\{a, b\}$. In

45

the set $\{a, b\}$ the elements a, b are not distinguished as 'first' and 'second': in fact as we have seen, $\{a, b\} = \{b, a\}$. In the pair (a, b), on the other hand, a is the *first component* and b is the *second component*. Again, the two components of an ordered pair may be the same: the two elements of a set are necessarily different.

The two components a, b of the ordered pair (a, b) which defines the position of a point P in the plane are called the *Cartesian co-ordinates* of P (after the French philosopher René Descartes, 1596–1650).

Example 1. Write the set $S = \{(x, y) \mid x \in N, y \in N, 5x + 4y = 68\}$ in roster notation. (Note that this is a re-statement of question 7, p. 43.)

The possible values of $x \in N$ are 1, 2, 3, 4, 5, . . .; but the logical possibilities exclude the odd positive integers for, if x is odd, $5x$ is odd, $68 - 5x$ is odd, and so is not a multiple of 4.

By trial $x = 4, y = 12; x = 8, y = 7; x = 12, y = 2$ are solutions: and they are the only solutions in positive integers, since any even value for x greater than 12 gives a negative value for y.

Hence $S = \{(4, 12), (8, 7), (12, 2)\}$.

4.2 Product sets

Example 2. Let $A = \{a, b\}$, $B = \{\alpha, \beta, \gamma\}$. Then, by choosing any element of A as the first component, and any element of B as the second, we can form a number of ordered pairs, *e.g.*, $(a, \alpha), (b, \gamma) \ldots$; how many are there altogether? The set of all such ordered pairs is called the *product set* of A and B.

Definition

If A and B are two sets, the product set of A and B is the set of all ordered pairs (a, b) such that $a \in A, b \in B$.

Notation

The product set of A and B is written $A \times B$ (read as 'A cross B'). In the above example:
$$A \times B = \{(a, \alpha), (a, \beta), (a, \gamma), (b, \alpha), (b, \beta), (b, \gamma)\}$$
$A \times B$ is sometimes referred to as the *Cartesian product* of A and B.

Note that A and B may in fact be the same set, as is the case with the points in a plane, whose Cartesian co-ordinates are all elements of the product set $R \times R$.

Example 3. Let $P = \{a, b, c\}$, $Q = \{b, c, d\}$, $R = \{a, c, d\}$. Then:

(a) $P \times (Q \cap R) = \{a, b, c\} \times \{c, d\}$
$= \{(a, c), (a, d), (b, c), (b, d), (c, c), (c, d)\}$

(b) $R \times (P \triangle Q) = \{a, c, d\} \times \{a, d\}$
$= \{(a, a), (a, d), (c, a), (c, d), (d, a), (d, d)\}$

(c) $(P - Q) \times (P - R) = \{a\} \times \{b\} = \{(a, b)\}$

Exercise 4a

1. Show by means of a counter example that the operation in sets denoted by $\times$ is not commutative.
2. If $A = \{a, b\}$, $B = \{b\}$, $C = \{b, c\}$, find $A \times (B \cap C)$ and $(A \times B) \cap (A \times C)$. What does the result suggest?
3. For the sets A, B, C defined in question 2, examine whether $\times$ is distributive over $\cup$.
4. If the universal set $\mathscr{E} = \{1, 2, 3\}$, $S_1 = \{1, 2\}$, $S_2 = \{1, 3\}$, $S_3 = \{2, 3\}$, write in roster notation the sets:
   (a) $S_1 \times S_2$
   (b) $S_1 \times S_3$
   (c) $(S_1 \times S_2) \triangle (S_1 \times S_3)$
   (d) $S_1' \times S_2$
   (e) $S_1 \times \emptyset$

Extension of product sets

The Cartesian product $A \times B \times C$ of three sets A, B, C is the set of all ordered triples (a, b, c), where $a \in A$, $b \in B$, $c \in C$.

Example 4. Let $P = \{$Arthur, Bertie$\}$; $Q = \{$Charles, David$\}$; and $R = \{$Enfield, Filey$\}$. Then $P \times Q \times R = \{(A, C, E), (A, C, F), (A, D, E), (A, D, F), (B, C, E), (B, C, F), (B, D, E), (B, D, F)\}$, writing the initial letters of the names for short. If P is a set of first Christian names, Q of second Christian names, and R of surnames, then $P \times Q \times R$ is the set of all eight names that can be formed. The probability of hitting by chance upon any given combination is $\frac{1}{8}$.

Cartesian products can be generalized. Where there are n sets $P_1, P_2, P_3, \ldots P_n$, the product set $P_1 \times P_2 \times P_3 \times \ldots \times P_n$ consists of all ordered n-tuples $(p_1, p_2, p_3, \ldots p_n)$ where $p_1 \in P_1$, $p_2 \in P_2$, $p_3 \in P_3, \ldots p_n \in P_n$.

4.3   The graph of a product set

Example 5. Consider again the sets $A = \{a, b\}$, $B = \{\alpha, \beta, \gamma\}$, of Example 2. The product set $A \times B$ may be illustrated by a lattice of six points as shown in Fig. 4.2. Why should these points not be joined by line segments?

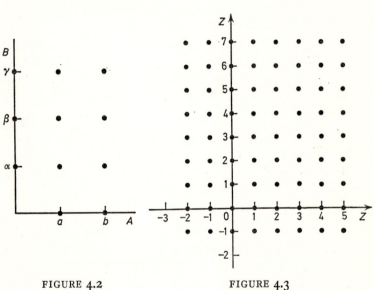

FIGURE 4.2   FIGURE 4.3

Example 6. Let $P = \{a \mid a \in Z, a > -3\}$, $Q = \{b \mid b \in Z, b > -2\}$. Again the graph of $P \times Q$ is a lattice, but it consists of an infinite number of points (Fig. 4.3). Here $P \times Q = \{(a, b) \mid a, b \in Z, a > -3, b > -2\}$.

Example 7. Let $P = \{x \mid x \in R, 1 < x < 2\}$, $Q = \{y \mid y \in R, -3 < y < -1\}$. Here the graph is the set of points within the shaded rectangle (Fig. 4.4). The points of the perimeter of the rectangle are not included: why?

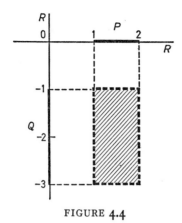

FIGURE 4.4

This subject is developed further in the book 'Relations and Functions' in this series.

Exercise 4b

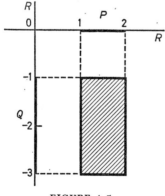

FIGURE 4.5

1. Figure 4.5 is the graph of a product set $P \times Q$ in which the points of the perimeter are included. Write down the sets P and Q.
2. Write out the product set $P \times P \times P$ where $P = \{p, f\}$.
3. For each of the following definitions of two sets A, B draw the graph of $A \times B$:
   (a) $A = \{x \mid x = 2a, a \in N, a < 5\}$, $B = \{1, 2, 3\}$

(b) $A = \{$April, Beth, Chloe, Denise$\}$
      $B = \{$Lamont, Martin, Nelson, Oates$\}$
(c) $A = \{x \mid x \in R,\ x < 3\}$, $B = \{y \mid y \in R,\ y > -3\}$

4.4   Tree diagrams

In certain types of problem the solution depends upon an analysis of the possibilities that can arise, and it is then important to ensure that none of these logical possibilities is omitted from consideration. 'Tree' diagrams (so called from their branching construction) are an aid. The following examples illustrate the use of these diagrams in connection with product sets.

Example 8. Consider the sets $P = \{$Arthur, Bertie$\}$; $Q = \{$Charles, David$\}$; $R = \{$Enfield, Filey$\}$ and the product set $P \times Q \times R$ (see Example 4, p. 47). With each of the two first names in set P may be associated each of the two second names in set Q, giving four combinations of Christian names in proper order. Then with each of these may be associated each of the two surnames, giving eight possible full names. The tree diagram in Fig. 4.6 enables these to be read off immediately (initials only are shown).

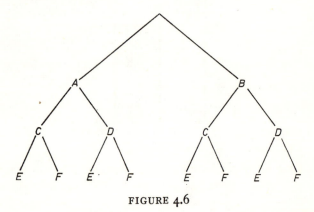

FIGURE 4.6

Example 9. If $X = \{\alpha,\ \beta,\ \gamma\}$, $Y = \{a,\ b,\ c\}$, $Z = \{1,\ 2,\ 3\}$, the product set $X \times Y \times Z$ has 27 ordered triples: these are given by the tree diagram of Fig. 4.7.

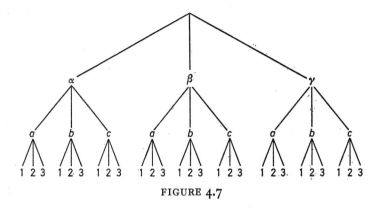

FIGURE 4.7

Example 10. A coin is tossed four times: how many combinations of heads (H) and/or tails (T) are possible, and what proportion of the total possibilities will show three heads and one tail?
The tree diagram is shown in Fig. 4.8.

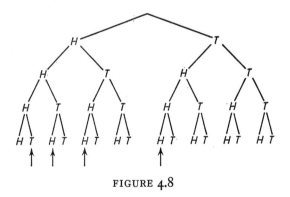

FIGURE 4.8

There are 16 possibilities, of which one-quarter have three heads and one tail (*HHHT, HHTH, HTHH, THHH*): they are indicated by arrows in Fig. 4.8.

Exercises on Chapter 4

1. Distinguish between (x, x) and (x, x, x).
2. If P, Q are two non-empty sets, and if $P \times Q = Q \times P$, prove that $P = Q$.

3. State and prove the converse of the theorem of the previous question: If P, Q are non-empty equal sets, then $P \times Q = Q \times P$.

4. Draw the graph of the product set $A \times B$ where:
$$A = \{x \mid x = 2a,\ a \in N,\ a < 5\}$$
$$B = \{y \mid y = (2a - 1),\ a \in N,\ a < 6\}$$

5. If $P \subset X$ and $Q \subset Y$, prove that $(P \times Q) \subset (X \times Y)$.
[Hint: begin by considering an element $(p,\ q)$ of $P \times Q$.]

6. Let $S = P \cap Q$, where $P = \{1,\ 2,\ 3,\ 4\}$, $Q = \{3,\ 4,\ 5\}$. Verify that:
(a) $S \times S = (P \times Q) \cap (Q \times P)$
(b) $S \times S = (P \times P) \cap (Q \times Q)$

7. Write the set $S = \{(p,\ q) \mid p \in Z,\ q \in Z,\ p^2 + q^2 = 169\}$ in roster notation, and draw its graph.

8. Let S be the set $\{(p,\ q) \mid p \in Z,\ q \in Z,\ p^2 + q^2 < 25\}$. Draw the graph of S.

9. A three-course lunch offers soup or fruit juice for the first course; joint, chicken or fish for the second; and fruit pudding, apple pie or coffee and biscuits for the third. Draw a tree diagram showing all the possible choices of a three-course lunch. In how many of these is fish chosen? In how many is soup followed by chicken?

10. Four roads $(r_1,\ r_2,\ r_3,\ r_4)$ lead from A to B; two $(s_1,\ s_2)$ from B to C and three $(t_1,\ t_2,\ t_3)$ from C to D. Draw a tree diagram showing all the possible routes from A to D through B and C. On one occasion r_3 was closed for repairs; on another occasion s_2 was closed. Which of these closures had the greater effect upon the number of available routes from A to D through B and C?

5 | PARTITIONING: COUNTING

5.1 Partitions

Example 1. Let $S = \{a, b, c, d, e\}$, and consider the following three sets of subsets of S:

(*i*) $\{\{a\}, \{a, b\}, \{c, d\}, \{e\}\}$
(*ii*) $\{\{a, b\}, \{c, e\}\}$
(*iii*) $\{\{a, e\}, \{b, c, d\}\}$

There are two features which distinguish (*iii*) from the other two sets of subsets. First, in (*iii*) no element of S appears in two different subsets: that is, the subsets are disjoint. This is not true in (*i*) where a is an element of two subsets. Second, in (*iii*) every element of S is included in some subset. This is not true in (*ii*) where d does not belong to any subset.

(*iii*) is an example of a partitioning of S.

Definition

A set S is said to be *partitioned* when it is divided up into a set P of disjoint subsets such that every element of S is an element of some one subset in P.

Example 2. Partition the set $S = \{$Ford, Simca, Hillman, Renault, Sunbeam, Fiat, Honda, Rover$\}$ in three different ways. If:

$P_1 = \{\{$Ford, Hillman, Sunbeam, Rover$\}$, $\{$Simca, Renault, Fiat, Honda$\}\}$

$P_2 = \{\{$Ford, Fiat$\}$, $\{$Simca, Sunbeam$\}$, $\{$Hillman, Honda$\}$, $\{$Renault, Rover$\}\}$

$P_3 = \{\{$Ford, Simca$\}$, $\{$Hillman, Renault, Sunbeam$\}$, $\{$Fiat, Honda, Rover$\}\}$

it is clear that P_1, P_2, P_3 are partitions of S, since in each case every element of S appears as an element of one and only one subset.

Write down three further partitions P_4, P_5, P_6 of S.

The process of partitioning a set is essentially one of *classification*. This is apparent in P_1 above: it consists of two subsets, the first containing British cars, and the second, foreign. Every car is either British or foreign, and no car is both: so according to the definition this classification provides a partition of S.

Again, in P_2 the classification is according to the initial letter of the name of the car. Here again both conditions for a partition are necessarily satisfied.

The principle of classification in P_3 is not obvious. This does not matter: the important point is that both conditions for a partition are once more met.

Exercise 5a

1. Let $S = \{n \mid n \in N, n < 10\}$. State which of the following sets of subsets of S are partitions of S:
    (*a*) $A = \{\{2, 4, 6, 8\}, \{1, 3, 5, 7, 9\}\}$
    (*b*) $B = \{\{2, 4, 8\}, \{3, 6, 9\}, \{5\}, \{7\}\}$
    (*c*) $C = \{\{1, 2, 3\}, \{3, 4, 5\}, \{5, 6, 7\}, \{7, 8, 9\}\}$
    (*d*) $D = \{\{1, 4, 6, 8, 9\}, \{2, 3, 5, 7\}\}$
    (*e*) $E = \{\{1\}, \{2\}, \{3\}, \{4\}, \{5\}, \{6\}, \{7\}, \{8\}, \{9\}\}$
    (*f*) $F = \{\{1, 4, 9\}, \{2, 3, 5, 6, 7, 8\}\}$
    What is the principle of classification in A, in D and in F?
2. If Z^+ is the set of positive integers, and Z^- is the set of negative integers, is $A = \{Z^+, Z^-\}$ a partition of Z? If you think it is not, make the necessary correction to A.
3. Partition the members of your form, classifying them according to the initial letter of their surname. Then partition the same set according to the months in which their birthdays fall.

Example 3. Refer again to question 1 above, and consider A and F. You have probably decided that these are true partitions of S. A was formed by putting together the even integers of S; F was formed by putting together the perfect squares in S. For reference:

$$S = \{1, 2, 3, \ldots 9\}$$
$$A = \{\{2, 4, 6, 8\}, \{1, 3, 5, 7, 9\}\}$$
$$F = \{\{1, 4, 9\}, \{2, 3, 5, 6, 7, 8\}\}$$

We may write down the intersection of each subset in A with each subset in F, thus:

$$\{2, 4, 6, 8\} \quad \cap \{1, 4, 9\} \qquad = \{4\}$$
$$\{1, 3, 5, 7, 9\} \cap \{1, 4, 9\} \qquad = \{1, 9\}$$
$$\{2, 4, 6, 8\} \quad \cap \{2, 3, 5, 6, 7, 8\} = \{2, 6, 8\}$$
$$\{1, 3, 5, 7, 9\} \cap \{2, 3, 5, 6, 7, 8\} = \{3, 5, 7\}$$

Now examine $G = \{\{4\}, \{1, 9\}, \{2, 6, 8\}, \{3, 5, 7\}\}$. G is another partition of S (check this). In it, not only have the even numbers been classified separately, and the perfect squares, but the even squares and the odd squares are classified separately. In other words, G follows the rules of classification of both A and F. G is called a *cross-partition* of A and F: it is a more detailed classification than either A or F.

Example 4. Blood obtained from donors is classified in two ways: (*i*) into four groups AB, A, B, O; (*ii*) into two groups, Rhesus + and Rhesus –. Each of these is a partition of the set of donations; and their cross-partition gives the eight blood types $AB +$, $AB -$, $A +$, $A -$, $B +$, $B -$, $O +$, $O -$.

Exercise 5b

1. Let $S = \{a, b, c, d, e\}$; $A = \{\{a, b\}, \{c, d, e\}\}$; $B = \{\{a, c, e\}, \{b, d\}\}$; $C = \{\{a, d\}, \{b, c, e\}\}$. Verify that A, B, C are partitions of S. Write down the cross-partitions of:
   (*a*) A and B (*b*) B and C (*c*) A and C
   What do you notice about the three cross-partitions?
2. Let $S = \{F, G, H, \mathcal{J}, K, L\}$, $A = \{\{F, H, K\}, \{G, \mathcal{J}, L\}\}$, $B = \{\{F, G\}, \{H, \mathcal{J}\}, \{K, L\}\}$. Are A, B partitions of S? If so, write down the cross-partition of A and B.

5.2 Counting the elements of a set

Notation

The number of elements in a finite set S is denoted by $n(S)$. A set S having r elements is said to be *of order r*.

Example 5. If $S_1 = \{x \mid x$ is an English consonant$\}$, then $n(S_1) = 21$. If $S_2 = \{x \mid x$ is a day of the week$\}$, then $n(S_2) = 7$.
   Note that $n(\emptyset) = 0$, but $n(\{0\}) = 1$.

Theorem I

Let S_1, S_2, S_3, . . . S_n be a set of subsets of S forming a partition of S: then $n(S) = n(S_1) + n(S_2) + n(S_3) + \ldots + n(S_n)$.

This follows since S is partitioned, and thus each element of S is contained once and only once in the partition.

Example 6. In Example 3 (p. 54), Theorem I is illustrated by the set S and the three partitions A, F, G. Note that:

$n(S) = 9$, $n(A) = 4 + 5 = 9$, $n(F) = 3 + 6 = 9$, $n(G) = 1 + 2 + 3 + 3 = 9$.

The following special case of Theorem I is important.

Theorem IA

If P, Q are two disjoint sets, then $n(P \cup Q) = n(P) + n(Q)$.

Note that this is a statement about numbers, hence the addition sign is appropriate. Distinguish carefully between $A \cup B$ and $n(A) + n(B)$. The first is a set produced by operating in sets; the second is a number obtained by operating in numbers.

Theorem II

If P, Q are any two sets, then $n(P \cup Q) = n(P) + n(Q) - n(P \cap Q)$.

Proof. P can be partitioned into two subsets $P \cap Q'$ (shaded horizontally in Fig. 5.1) and $P \cap Q$ (shaded vertically). Hence, by Theorem I:

$$n(P) = n(P \cap Q') + n(P \cap Q) \tag{5.1}$$

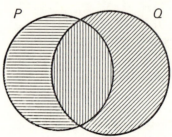

FIGURE 5.1

Similarly Q can be partitioned into $Q \cap P'$ (shaded obliquely) and $P \cap Q$, thus:

$$n(Q) = n(Q \cap P') + n(P \cap Q) \tag{5.2}$$

From (5.1) and (5.2)

$$n(P) + n(Q) = n(P \cap Q') + n(Q \cap P') + 2n(P \cap Q)$$
$$n(P) + n(Q) - n(P \cap Q) = n(P \cap Q') + n(Q \cap P') + n(P \cap Q)$$
$$= n(P \cup Q) \quad \text{by (Theorem I)}$$

Example 7. If $n(X) = 25$, $n(Y) = 17$, $n(X \cap Y) = 4$, then from Theorem II we have $n(X \cup Y) = 38$.

Theorem III

If P, Q, R are any three sets, then:

$$n(P \cup Q \cup R) = n(P) + n(Q) + n(R) - n(P \cap Q) - n(Q \cap R) - n(R \cap P) + n(P \cap Q \cap R)$$

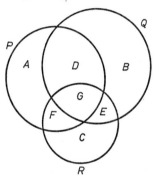

FIGURE 5.2

This may be proved on the lines of Theorem II, or it may be shown as follows. In Fig. 5.2, A, D, F, G are subsets of P which partition P, hence:

$$n(P) = n(A) + n(D) + n(F) + n(G) \quad \text{(Theorem I)}$$

Similarly:
$$n(Q) = n(B) + n(E) + n(G) + n(D)$$
$$n(R) = n(C) + n(F) + n(G) + n(E)$$

Thus:

$$\begin{aligned}
n(P) + n(Q) + n(R) &= [n(A) + n(D) + n(F) + n(G) + n(B) + n(E) + n(C)] \\
&\quad + [n(D) + n(G)] + [n(E) + n(G)] + [n(F) + n(G)] \\
&\quad - n(G) \\
&= n(P \cup Q \cup R) + n(P \cap Q) + n(Q \cap R) + n(R \cap P) \\
&\quad - n(P \cap Q \cap R)
\end{aligned}$$

whence the theorem follows.

Example 8. In an investigation into the pets (birds, cats or dogs) preferred by a group of 75 children the following information was obtained: 37 liked birds, 33 liked cats, 40 liked dogs; 16 liked both birds and cats, 11 liked both cats and dogs, 12 liked both birds and dogs. How many liked all three, assuming that no-one disliked all three?

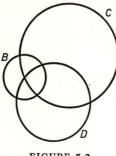

FIGURE 5.3

In Fig. 5.3, let B represent the set of those who liked birds; C, the set of those who liked cats; and D, the set of those who liked dogs. We are given $n(B) = 37$, $n(C) = 33$, $n(D) = 40$, $n(B \cap C) = 16$, $n(C \cap D) = 11$, $n(B \cap D) = 12$. Also, $n(B \cup C \cup D) = 75$. Whence, by Theorem III:

$$n(B \cap C \cap D)$$
$$= 75 - (37 + 33 + 40 - 16 - 11 - 12)$$
$$= 4$$

This example lends itself to a straightforward application of Theorem III. This, however, is not always the case, and it is often simpler to work from a Venn diagram without reference to a theorem.

Example 9. Weather records kept for a certain March showed that the month had no warm, calm, dry days. Of the 31 days, 7 were wet and cold but not windy, 4 were wet and windy but not cold, 8 were cold and windy but dry. If 16 days were windy, 22 were cold and 2 were wet but not cold or windy, (*i*) how many cold, wet, windy days were there, and (*ii*) how many days were cold but calm and dry?

In Fig. 5.4 the numbers 7, 4, 8, 2 can be filled in immediately. Denote by x, y, z the numbers of elements in the remaining regions:

$$x + y + 12 = 16 \text{ (the total number of windy days)}$$
$$x + z + 15 = 22 \text{ (the total number of cold days)}$$
$$x + y + z + 21 = 31 \text{ (the total number of days in the month)}$$

These equations lead to the result $x = 1$, $z = 6$. So there was 1 cold, wet, windy day, and there were 6 cold, calm, dry days.

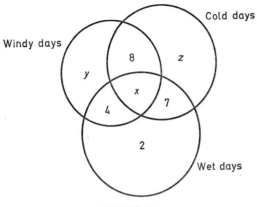

FIGURE 5.4

Exercise 5c

1. If $(P \cap Q) = \emptyset$, $n(P) = 5$, $n(P \cup Q) = 9$, find $n(Q)$.
2. If $n(P \cap Q) = 6$, $n(P) = 8$, $n(P \cup Q) = 11$, find $n(Q)$.
3. If P is partitioned into subsets P_1, P_2, P_3, and $n(P) = 14$, $n(P_1) = 4$, $n(P_2) = 7$, find $n(P_3)$.
4. Given $n(P \cup Q \cup R) = 20$, $n(P) = n(Q) = n(R) = 9$, $n(P \cap Q) = n(Q \cap R) = n(R \cap P) = 3$, find $n(P \cap Q \cap R)$.
5. Show that Theorem I is a special case of Theorem III.
6. If a set P is of order 5, and a set Q is of order 7, what is the most that can be said about the order of these sets:
   (a) $P \cup Q$ (b) $P \cap Q$

5.3 Permutations of the elements of a set

The problem of counting, that is, the search for an answer to the question 'How many?' often involves more than a simple enumeration. Counting the number of people in a room is straightforward; but counting the number of ways in which the same people may be arranged at table is rather more difficult. If there are ten guests apart from the host and hostess, they may be arranged in ten places in over $3\frac{1}{2}$ million different ways!

The problem is that of counting the number of different orders in which the elements of a set S may be written: that is, the number of *permutations* of the elements of S.

C

Example 10. How many permutations are there of the elements of $S = \{a, b, c\}$?

In this simple case the permutations may be written out:

$$a, b, c \qquad b, a, c \qquad c, a, b$$
$$a, c, b \qquad b, c, a \qquad c, b, a$$

That there are no more than these six arrangements may be shown thus: element a can occupy any one of three positions; then b can occupy either of the remaining two positions; and c must then take the remaining place. So along with each of 3 positions of a there are 2 of b, making 6 permutations in all.

Consider the case of the ten guests mentioned earlier. The first may be put in any one of 10 places, then the second in any one of the remaining 9, the third in any one of the remaining 8, and so on. The last guest must take the one remaining place. Hence the total number of permutations is $10.9.8.7.6.5.4.3.2.1 = 3\,628\,800$, a result which is perhaps unexpectedly large.

In general we say that, if S has n elements they may be arranged (or *permuted*) in $n(n-1)\,(n-2)\,(n-3)\,\ldots\,4.3.2.1$ ways.

Notation

Products such as this, where a positive integer n is multiplied by all the positive integers less than n, are very common, and are denoted by $n!$ (read 'n factorial') The symbol $\underline{|n}$ is also used. Note that $a.b!$ is to be understood as $a(b!)$.

There are $n!$ permutations of the n elements of a set S, that is, of n things all different.

Example 11. If $S = \{a, e, i, o, u\}$, then $n(S) = 5$ and the number of permutations of the English vowels is $5! = 5.4.3.2.1 = 120$. In how many of these arrangements does e come first?

Exercise 5d

1. Write down the values of:

    (*a*) 6!   (*b*) 8!   (*c*) 1!

2. Evaluate:

    (*a*) $\dfrac{7!}{6!}$ (*b*) $\dfrac{8!}{5!3!}$ (*c*) $\dfrac{n!}{(n-2)!}$

3. Express in factorial notation:
   (a) $5.4!$ (b) $(a-2).(a-3).(a-4)!$
4. Prove that $5! + 4! = 24.3!$
5. Simplify $6! + 5! + 4!$, and prove that in general:
   $n! + (n-1)! + (n-2)! = n^2.(n-2)!$

5.4 Permutations of the elements of subsets

Example 12. Consider the set $S = \{p, q, r, s\}$ and all those of its subsets which are of order 3, *i.e.*, have 3 elements. In how many ways can the elements of these subsets be arranged among themselves?

There are four subsets of S of order 3:

$$\{p, q, r\}, \quad \{p, q, s\}, \quad \{p, r, s\}, \quad \{q, r, s\}$$

The elements of each of these subsets may be arranged in $3! = 6$ ways: so there are 24 permutations of the elements of the four subsets. This may be generalized.

Theorem IV

If a set S is of order n, the number of permutations of the elements of those of its subsets which are of order r is

$$\frac{n!}{(n-r)!}$$

Proof. The first element of a subset S_1 may be any one of the n elements of S: the second may be any one of the remaining $(n-1)$ elements of S, and so on. The rth element of S_1 may be chosen from the remaining $(n-r+1)$ elements of S. Hence the total number of permutations is $n(n-1)(n-2)(n-3) \ldots (n-r+1)$.

This is not a factorial, but it may be expressed in factorial notation thus:

$$n(n-1)(n-2)(n-3) \ldots (n-r+1)$$
$$= n(n-1)(n-2) \ldots (n-r+1).\frac{(n-r)!}{(n-r)!}$$
$$= \frac{n!}{(n-r)!}$$

When $n = 4$, $r = 3$, the number of permutations is $\frac{4!}{1!} = 24$ as was found in Example 12.

Example 13. In how many ways can 5 people be seated in a railway compartment with seats for 8 passengers?

The 5 occupied seats form a subset of the 8 available seats. Hence Theorem IV applies, and the number of permutations is:

$$\frac{8!}{3!} = 8.7.6.5.4 = 6720$$

Exercise 5e

1. Write in factorial notation:
   (a) $11.10.9.8$
   (b) $x(x-1)(x-2)(x-3) \ldots (x-y)$ (c) $\dfrac{8.7.6.5.4}{5.4.3.2.1}$
2. A set S is of order 6. Write down the number of permutations of the elements of those subsets of S that are of order 3.
3. How many 6-digit numbers can be formed from the numbers 1 to 9, if no digit is repeated in any number?
4. If $n(S) = 8$, $A \subset S$, $n(A) = 5$, find the number of arrangements of the elements of A.

5.5 Number of subsets of a given order

Theorem V

A set S of order n has $\dfrac{n!}{r!(n-r)!}$ subsets of order r.

Proof. Let S have x subsets of order r. The elements of each of these subsets may be permuted in $r!$ ways and the total number of permutations of the elements of all these subsets is:

$$\frac{n!}{(n-r)!} \quad \text{(Theorem IV)}$$

Hence: $\qquad\qquad r!x = \dfrac{n!}{(n-r)!}$

$$x = \frac{n!}{r!(n-r)!}$$

Example 14. How many committees of six members can be elected from nine nominees?

Here we are concerned only with *selections*: the arranging among themselves of the six elected members is irrelevant.

Hence the possible number of committees is the number of subsets of order 6 belonging to a set of order 9. This is:

$$\frac{9!}{6!3!} = \frac{9.8.7}{3.2.1} = 84$$

Notation

The expression in Theorem V occurs frequently and is denoted by $\binom{n}{r}$, or sometimes by ^nC_r or C_r^n.

$$\binom{n}{r} = \frac{n!}{r!(n-r)!}$$

Exercise 5f

1. Evaluate:

   (a) $\binom{3}{2}$ (b) $\binom{10}{6}$ (c) $\binom{n}{n-2}$

2. Show that $\binom{n}{r} = \binom{n}{n-r}$

3. Show that:

   (a) $\binom{10}{4} = \frac{5}{6} \cdot \binom{10}{5}$

   (b) $\binom{8}{3} = \frac{2}{3} \cdot \binom{9}{3}$

   (c) $\binom{12}{5} = \binom{11}{5} + \binom{11}{4}$

Note

We have defined $n!$ for $n \in N$, which excludes the case where $n = 0$. But if in the formula $\binom{n}{r} = \frac{n!}{r!(n-r)!}$ we put $r = 0$, we have $\binom{n}{0} = \frac{n!}{0!n!} = \frac{1}{0!}$; and this is 1, being the number of ways of selecting no object from n. The single way of doing this is to reject all n objects. For this reason we assign the value 1 to 0!

5.6   Number of partitions with subsets of given order

Example 15. If $S = \{a, b, c, d\}$, how many partitions of S can be formed, each having two subsets, one with three elements and the other with one?

We may answer this by listing the possibilities:

$\{\{a, b, c\}, \{d\}\}$ $\{\{a, b, d\}, \{c\}\}$ $\{\{a, c, d\}, \{b\}\}$ $\{\{b, c, d\}, \{a\}\}$

But we do not have to. The question is: 'In how many ways can one element be chosen from S for the second subset?' The answer is, of course, 4; or, by Theorem V, it is $\binom{4}{1} = \dfrac{4!}{1!3!} = 4$.

Example 16. If $S - \{a, b, c, d\}$, how many partitions of S can be formed each having three subsets of order 2, 1, 1, respectively?

For the first subset, 2 elements can be chosen from 4 in $\binom{4}{2} = 6$ ways: then the remaining elements will form the other two subsets.

One of these six possible partitions is $\{\{a, b\}, \{c\}, \{d\}\}$: write down the other five.

Ordered partitions

Since a partition is a set, the order in which its elements, *i.e.*, the subsets which comprise it, are written is unimportant. The partitions $P_1 = \{\{a, b, c\}, \{d\}\}$ and $P_2 = \{\{d\}, \{a, b, c\}\}$ of $S = \{a, b, c, d\}$ are identical.

However, it is sometimes convenient to regard P_1 and P_2 as being different. We say that, while they are the same partition, they are different *ordered* partitions.

Definition. An ordered partition is a partition whose subsets are written in a specified order. Distinguish between 'the order in which subsets are written' and 'the order of a subset', which is the number of elements it contains.

In Example 16, if the question had been 'How many ordered partitions of $S = \{a, b, c, d\}$ can be formed each having three subsets with 2, 1, 1 elements, respectively?', we should have had to include $\{\{a, b\}, \{d\}, \{c\}\}$ as well as $\{\{a, b\}, \{c\}, \{d\}\}$, since both these partitions preserve the order 2, 1, 1. There are 12 ordered partitions in this case.

The question may be answered directly thus: for the last subset, one element may be chosen from 4 in $\dfrac{4!}{1!3!}$ ways. Then for the second subset, one element may be chosen from the remaining 3 in $\dfrac{3!}{1!2!}$ ways. The remaining elements must go into the first subset.

There are, therefore, altogether $\dfrac{3!}{1!2!} \cdot \dfrac{4!}{1!3!} = \dfrac{4!}{2!1!1!} = 12$ different ordered partitions of S satisfying the given condition. This may be generalized.

Theorem VI

If S is a set of order n, partitioned into k subsets of order a_1, a_2, a_3, . . . a_k, respectively, $(a_1 + a_2 + a_3 + . . . + a_k = n)$, then the number of such ordered partitions is:

$$\frac{n!}{a_1! a_2! a_3! \ldots a_k!}$$

Example 17. If $n(S) = 7$, and S is partitioned into subsets having 3, 2, 1, 1 elements, then the number of such ordered partitions is:

$$\frac{7!}{3!2!1!1!} = \frac{7.6.5.4}{2.1} = 420$$

Example 18. How many arrangements are possible of the letters of the word *rigging*?

If the seven letters were all different, the number of arrangements would be 7! But the 3! permutations of the letters g among themselves, and the 2! permutations of the letters i among themselves do not give new permutations: hence the number of permutations is $\dfrac{7!}{3!2!} = 420$.

Observe that the last two examples are mathematically equivalent.

Exercise 5g

1. If $S = \{n \mid n \in N, n < 10\}$, how many ordered partitions of S are possible if all subsets are to be of order 3?

2. How many different arrangements are possible of the letters 'SOS'? Write them out.
3. How many different arrangements are possible of the letters of the words:
    (a) *Passage*   (b) *Passages*
4. If a set S is of order n, how many partitions of S are possible if each subset is of order 1? How many *ordered* partitions are possible in this case?

5.7 The binomial expansion

Consider the expression $(1 + x)^n$, $(n \in N)$. It is the product of n factors, each being $(1 + x)$:

$$(1 + x)^n = (1 + x)(1 + x)(1 + x) \ldots (1 + x) \qquad (n \text{ factors})$$

For any given value of $n \in N$ this product may be obtained by multiplying each term of each factor by each term of every other factor, *e.g.*:

$$
\begin{aligned}
(1 + x)^2 &= (1 + x)(1 + x) \\
&= (1.1 + 1.x) + (x.1 + x.x) \\
&= 1 + 2x + x^2 \\
(1 + x)^3 &= (1 + x)(1 + x)(1 + x) \\
&= (1.1.1 + 1.1.x) + (1.x.1 + 1.x.x) + (x.1.1 + x.1.x) \\
&\quad + (x.x.1 + x.x.x) \\
&= 1 + 3x + 3x^2 + x^3
\end{aligned}
$$

In the expansion of $(1 + x)^n$ there will be a term in $x^r (r \in N, r < n)$, and it will be the sum of all the terms in which x^r is obtained in all possible ways. But there are $\binom{n}{r}$ ways of choosing x r times out of n (Theorem V): hence the term in x^r will be $\binom{n}{r} x^r$.

Theorem VII

$$(1 + x)^n = \binom{n}{0} x^0 + \binom{n}{1} x^1 + \binom{n}{2} x^2 + \ldots + \binom{n}{r} x^r + \ldots + \binom{n}{n} x^n$$

This is the simplest case of the *binomial expansion* for a positive integral index (*i.e.*, $n \in N$). The general case where $n \in R$ is beyond the scope of this book.

To $\binom{n}{0}$ we have assigned the value 1: $\binom{n}{n}$ is also 1—why? By

Theorem V, $\binom{n}{1} = n$, $\binom{n}{2} = \dfrac{n(n-1)}{2!}$, $\binom{n}{3} = \dfrac{n(n-1)\,(n-2)}{3!}$, and so

on. The general term, $\binom{n}{r}x^r$, is $\dfrac{n!}{r!\,(n-r)!}x^r$.

Example 19. The coefficient of x^3 in the expansion of $(1+x)^{11}$ is
$$\binom{11}{3} = \frac{11!}{3!\,8!} = \frac{11.10.9}{3.2.1} = 165.$$

Note that $\binom{11}{3} = \binom{11}{8}$. Why? It follows that the coefficients of x^3
and x^8 are identical. What other terms have the same coefficients as
x^5 and x^{10}?

Exercise 5h

1. Write out the expansions and evaluate the coefficients of:
   (a) $(1+x)^4$ (b) $(1+x)^5$
2. Evaluate the coefficients of:
   (a) x^3 (b) x^5
   in the expansion of $(1+x)^{10}$. What other term has the same
   coefficient as x^3? Has any other term the same cofficient as x^5?
   How many different coefficients have to be evaluated to obtain
   the complete expansion?
3. Find the cofficient of x^6 in the expansion of $(1+x)^{13}$. Has any
   other term the same coefficient as x^6?
4. Write down the coefficients of x^{n-3}, x^{n-2} and x^{n-1} in the
   expansion of $(1+x)^n$.

5.8 Number of subsets of a given set

Theorem VIII

A set S of order n has 2^n subsets (including $\emptyset$ and S).

First Proof. The total number of subsets will be the sum of the numbers of those that have 0, 1, 2, 3, . . n elements, respectively. But
the number of subsets of order r is $\binom{n}{r}$ (Theorem VI). Hence the
total possible number of subsets is:

$$\binom{n}{0} + \binom{n}{1} + \binom{n}{2} + \ldots + \binom{n}{r} + \ldots + \binom{n}{n}$$

If in the binomial expansion we put $x = 1$, we have:

$$2^n = \binom{n}{0} + \binom{n}{1} + \binom{n}{2} + \ldots + \binom{n}{r} + \ldots + \binom{n}{n}$$

Hence the total possible number of subsets of a set S of order n is 2^n.

Second Proof. The theorem may be proved using the principle of finite induction.

If $n = 1$, S has one element, say $S = \{a\}$. Then S has two subsets, S and $\emptyset$.

If $n = 2$, say $S = \{a, b\}$, there are $4 = 2^2$ subsets, $\{a, b\}$ $\{a\}$, $\{b\}$, $\emptyset$.

Similarly, if $n = 3$, there are $8 = 2^3$ subsets. Write them out for $S = \{a, b, c\}$.

Assume that if $n = r$ there are 2^r subsets. Then, when $n = r + 1$ the additional element of S may be included in each of these 2^r subsets or excluded from each of them: there are thus $2 . 2^r = 2^{r+1}$ subsets.

So if the result is true for $n = r$ it is true for $n = r + 1$. But it is true for $n = 1$, hence for $n = 2, 3, \ldots$ and so on for all $n \in N$.

Exercises on Chapter 5

1. Prove $\binom{n}{r} + 2\binom{n}{r-1} + \binom{n}{r-2} = \binom{n+2}{r}$

2. The universal set $\mathscr{E}$ is partitioned into four subsets. Three of them are $P' \cap Q$, $P \cap Q'$, $P \cap Q$: what is the fourth?

3. An investigation conducted into the reaction of 120 people to a proposal that baseball should be introduced into this country produced the following results:

	For	Against	Uncertain
Adults	10	18	2
Boys	23	15	12
Girls	24	6	10

Let M be the set of adults and boys; Y the set of young people; F the set of people who were for the proposal; A the set of those against the proposal. Find:

(a) $n(M \cap Y)$ (c) $n(F \cap Y')$
(b) $n(Y')$ (d) $n[(M - Y') \cap (F \cup A)]$

Write in terms of M, Y, F, A the sets of:

(e) girls who were against the proposal

(f) boys who were against or uncertain about the proposal

(g) adults and girls who were for the proposal

(h) adults who were uncertain about the proposal

4. Prove $\binom{n+1}{r} = \binom{n}{r} + \binom{n}{r-1}$.

5. 2500 people were questioned about their appreciation of orchestral music (O), piano music (P) and choral music (C). The returns showed that 674 appreciated all three, 910 O and P, 828 O and C, 1032 P and C, 321 only O, 286 only P, 384 only C. How many were unappreciative of any of the three?

6. Two sets P, Q are said to be *equivalent* if $n(P) = n(Q)$. If P, Q, R are three equivalent sets of which Q and R are disjoint, prove:
$$n(P \cup Q \cup R) = 3n(P) - n(P \cap Q) - n(P \cap R)$$

7. A postcard has a hole punched in one corner. A rectangle with the same dimensions as the card is drawn on paper. Let S be the set of ways in which the card may be placed in the rectangle (with either face showing) so that the hole appears in a different corner of the rectangle each time. Find $n(S)$.

   If the card were an equilateral triangle, what would $n(S)$ be?

   If the card were a parallelogram, what would $n(S)$ be?

8. How many even numbers exceeding 10000 can be formed from the digits 1, 3, 4, 5, 6 if no digit appears more than once in any number?

9. In how many ways can 18 different objects be packed in:

   (a) 3 parcels of 6 objects (b) 6 parcels of 3 objects

10. A student is to answer seven questions out of nine in an examination.

    (a) How many choices has he?

    (b) How many choices has he if he must answer the first three questions?

    (c) How many choices has he if he must answer at least three questions out of the first four? (AEB)

11. By writing $(2+x)^{10} = 2^{10}\left(1 + \dfrac{x}{2}\right)^{10}$, prove that the coefficient of x^7 in the expansion of $(2+x)^{10}$ is 960. What is the value of the middle term in the expansion of $(2+x)^{10}$ when $x = \frac{1}{2}$?

12. In a sixth form, 16 take physics, 17 chemistry, 15 biology, 5 physics and chemistry, 8 physics and biology, 3 take all three subjects, and 31 take at least one of them.

(*a*) How many take only chemistry?

(*b*) How many take only biology?

(*c*) How many take chemistry and biology?

6 | LAWS OF THE ALGEBRA OF SETS

6.1 The algebra of sets

The word 'algebra' doubtless suggests the familiar technique of examining the relations that exist among numbers, and the operations that may be performed in them. The word has, however, a wider meaning. Given any set of elements related in specified ways, and an operation or operations which may be performed in that set, we have the materials for constructing an algebra. This subject is developed in the book 'Mathematical Structures' in this series.

It is clear that we may legitimately speak of 'the algebra of sets'. In the course of our development of set theory so far we have studied various operations in sets and have introduced certain laws which these operations obey. We are now in a position to summarize these laws.

6.2 The laws of sets

These may conveniently be set out in five groups. (P, Q, R represent sets.)

A. Primary laws

1. $P \cup P = P$
2. $P \cap P = P$
3. $P \cup \emptyset = P$
4. $P \cap \mathscr{E} = P$
5. $P \cup \mathscr{E} = \mathscr{E}$
6. $P \cap \emptyset = \emptyset$

B. Associative laws

1. $(P \cup Q) \cup R = P \cup (Q \cup R)$
2. $(P \cap Q) \cap R = P \cap (Q \cap R)$

C. Commutative laws

1. $P \cup Q = Q \cup P$
2. $P \cap Q = Q \cap P$

D. *Distributive laws*

1. $P \cup (Q \cap R) = (P \cup Q) \cap (P \cup R)$
2. $P \cap (Q \cup R) = (P \cap Q) \cup (P \cap R)$

E. *Laws of complements*

1. $P \cup P' = \mathscr{E}$ 2. $P \cap P' = \emptyset$
3. $\emptyset' = \mathscr{E}$ 4. $\mathscr{E}' = \emptyset$
5. $(P \cup Q)' = P' \cap Q'$ 6. $(P \cap Q)' = P' \cup Q'$
7. $(P')' = P$

Note 1. Since the set operations of union and intersection are commutative (laws C), it follows that the distributive laws (*D*) may also be written:

$$(Q \cap R) \cup P = (Q \cup P) \cap (R \cup P)$$
$$(Q \cup R) \cap P = (Q \cap P) \cup (R \cap P)$$

That is to say, the union of sets is both left- and right-distributive over intersection: and the intersection of sets is both left- and right-distributive over union.

Note 2. Laws E5 and E6 are usually known as *DeMorgan's laws*. (Augustus DeMorgan, an English Mathematician, lived 1806–71: he first introduced the method of finite induction.)

Exercise 6a

With the help of Venn diagrams, verify laws A1–6 and E1–7.

6.3 The principle of duality

Observe that the six laws in group A occur in three pairs, and that either member of any pair may be derived from the other by two interchanges:

(*i*) of $\cup$ and $\cap$ (*ii*) of $\mathscr{E}$ and $\emptyset$

It will be seen that this applies in each group: the laws are in related pairs with the exception of E7, which does not admit of either of the interchanges (*i*) and (*ii*).

This pairing of the laws has two consequences. For one thing, they can more readily be remembered; for another, the pairing principle can be extended.

If any theorem relating to sets has been proved using the laws of sets, then a parallel proof may be set out with the interchanges (i) and (ii) made throughout. As we have seen, there will be a law to substantiate each step of the new proof, and hence the new conclusion will be valid. Thus any theorem relating to sets and their union or intersection is matched by another equally valid theorem obtained by making interchanges (i) and (ii) as appropriate. Where subsets are involved, $\subset$ and $\supset$ must also be interchanged.

Example 1. It may be proved (see Example 2) that for any two sets P, Q, $(P \cup Q) \cap (P \cup Q') = P$. Without further proof, it may be stated that $(P \cap Q) \cup (P \cap Q') = P$.

Each of these theorems is the *dual* of the other.

Definition

The *principle of duality* states that if any theorem is proved true, the dual of that theorem is also true.

Exercise 6b

1. Write the dual of each of the following theorems:
   (*a*) $(A \cap B') \cap B = \emptyset$
   (*b*) $(A \cup \mathscr{E}) \cap (A \cap \emptyset) = \emptyset$
   (*c*) If $A \cap B = \emptyset$, then $A' \cup B = A'$
   (*d*) If $(A \cap B') = A \cap B$, then $A = \emptyset$
   (*e*) $A \cup B = (A \cap B) \cup (A \cap B') \cup (A' \cap B)$
   (*f*) $(A \cup \emptyset) \cup (\mathscr{E} \cap A') = \mathscr{E}$
   (*g*) $(A \cap B) \subset (A \cup B)$
2. Verify each of the theorems of question 1 using Venn diagrams.

6.4 Set theorems

When it is required to prove a theorem in the algebra of sets, there are (so far) two methods open to us. It has already been stated that drawing Venn diagrams is not one of them. Venn diagrams may usefully illustrate the theorem, but they cannot prove it, any more than the drawing of a figure can establish a geometrical theorem.

Proof by a consideration of elements

This may be illustrated by some examples.

Example 2. Prove that, for two sets $P, Q, (P \cup Q) \cap (P \cup Q') = P$.
Remembering the definition of the equality of sets, we must prove:

(*i*) $[(P \cup Q) \cap (P \cup Q')] \subset P$
(*ii*) $P \subset [(P \cup Q) \cap (P \cup Q')]$

(*i*) Let $x \in [(P \cup Q) \cap (P \cup Q')]$.
Then $x \in (P \cup Q)$ and $x \in (P \cup Q')$ (definition of $\cap$)
$\Rightarrow$ ($x \in P$ or $x \in Q$) and ($x \in P$ or $x \in Q'$) (definition of $\cup$)
$\Rightarrow$ $x \in P$ or ($x \in Q$ and $x \in Q'$) (distributive law)
The latter alternative is impossible (definition of complement);
hence $x \in P$. Thus every element x of $(P \cup Q) \cap (P \cup Q')$
is an element of P, and $[(P \cup Q) \cap (P \cup Q')] \subset P$.
(*ii*) Let $x \in P$.
Then $x \in (P \cup Q)$ and $x \in (P \cup Q')$ (definition of $\cup$)
$\Rightarrow$ $x \in [(P \cup Q) \cap (P \cup Q')]$ (definition of $\cap$)
So every element x of P is an element of $[(P \cup Q) \cap (P \cup Q')]$;
hence $P \subset [(P \cup Q) \cap (P \cup Q')]$.
From (*i*) and (*ii*), $(P \cup Q) \cap (P \cup Q') = P$.

Example 3. Prove DeMorgan's laws (E5, E6).
To prove $(P \cup Q)' = P' \cap Q'$ (E5).

(*i*) Let $x \in (P \cup Q)'$.
Then $x \notin (P \cup Q)$
$\Rightarrow$ $x \notin P$ and $x \notin Q$
$\Rightarrow$ $x \in P'$ and $x \in Q'$
$\Rightarrow$ $x \in (P' \cap Q')$
Hence $(P \cup Q)' \subset (P' \cap Q')$.
(*ii*) Let $x \in P' \cap Q'$.
Then $x \in P'$ and $x \in Q'$
$\Rightarrow$ $x \notin P$ and $x \notin Q$
$\Rightarrow$ $x \notin (P \cup Q)$
$\Rightarrow$ $x \in (P \cup Q)'$
Hence $P' \cap Q' \subset (P \cup Q)'$.
From (*i*) and (*ii*), $(P \cup Q)' = P' \cap Q'$. Law E6 follows by the
principle of duality.

Exercise 6c

By a consideration of elements, prove the following theorems for all sets P, Q, R:

1. $P \cap (P \cup Q) = P$
2. $Q \cap (Q \cap R') = Q \cap R'$
3. $P \cap (P' \cup Q) = P \cap Q$
4. $P \cup [(P \cup Q) \cap (P' \cup Q')] = P \cup Q$

Proof using the laws of sets

It is important that in a proof no statement be made without the support of a definition (*e.g.*, one of the laws of Section 6.2), or a theorem already derived from these laws. In the following examples the references to the laws are self-explanatory.

Example 4. Prove $P \cap (P' \cup Q) = P \cap Q$ (*cf.* question 3 above).

$$
\begin{aligned}
\text{LHS} = P \cap (P' \cup Q) &= (P \cap P') \cup (P \cap Q) & \text{(D2)} \\
&= \emptyset \cup (P \cap Q) & \text{(E2)} \\
&= (P \cap Q) \cup \emptyset & \text{(C1)} \\
&= (P \cap Q) & \text{(A3)} \\
&= \text{RHS}
\end{aligned}
$$

The laws of sets do not include any reference to either the difference $(P - Q)$ or the symmetric difference $(P \triangle Q)$ of two sets P, Q. In fact:

$$P - Q = P \cap Q' \text{ (cf. Exercise 2d, question 3, p. 19)} \quad (6.1)$$
$$\text{and} \quad P \triangle Q = (P \cup Q) - (P \cap Q) \text{ (section 2.6)} \quad (6.2)$$

From (6.1), $P \triangle Q$ may be written $(P \cup Q) \cap (P \cap Q)'$; hence both difference and symmetric difference may be expressed in terms of union, intersection and complement.

Example 5. Prove $P - (Q - R) = (P - Q) \cup (P \cap R)$.

$$
\begin{aligned}
\text{LHS} = P - (Q - R) &= P \cap (Q \cap R')' & \text{(6.1 above)} \\
&= P \cap [Q' \cup (R')'] & \text{(E6)} \\
&= P \cap (Q' \cup R) & \text{(E7)} \\
&= (P \cap Q') \cup (P \cap R) & \text{(D2)} \\
&= (P - Q) \cup (P \cap R) & \text{(6.1)} \\
&= \text{RHS}
\end{aligned}
$$

Example 6. Prove $(P - Q) \cap (P - R) = P - (Q \cup R)$.

$$
\begin{aligned}
\text{LHS} = (P - Q) \cap (P - R) &= (P \cap Q') \cap (P \cap R') && (6.1) \\
&= (P \cap P) \cap (Q' \cap R') && (C2) \\
&= P \cap (Q' \cap R') && (A2) \\
&= P \cap (Q \cup R)' && (E5) \\
&= P - (Q \cup R) && (6.1) \\
&= \text{RHS}
\end{aligned}
$$

Example 7. Prove that intersection distributes over difference, *i.e.*:
$$P \cap (Q - R) = (P \cap Q) - (P \cap R).$$

$$
\begin{aligned}
\text{RHS} = (P \cap Q) - (P \cap R) &= (P \cap Q) \cap (P \cap R)' && (6.1) \\
&= (P \cap Q) \cap (P' \cup R') && (E6) \\
&= [(P \cap Q) \cap P'] \cup [(P \cap Q) \cap R'] && (D2) \\
&= [(P \cap P') \cap Q] \cup [(P \cap Q) \cap R'] && (C2) \\
&= (\emptyset \cap Q) \cup [(P \cap Q) \cap R'] && (E2) \\
&= \emptyset \cup [(P \cap Q) \cap R'] && (A6) \\
&= (P \cap Q) \cap R' && (A3) \\
&= P \cap (Q \cap R') && (B2) \\
&= P \cap (Q - R) && (6.1) \\
&= \text{LHS}
\end{aligned}
$$

Example 8. Prove that intersection distributes over symmetric difference, *i.e.*:
$$P \cap (Q \,\Delta\, R) = (P \cap Q) \,\Delta\, (P \cap R)$$

$$
\begin{aligned}
\text{RHS} = (P \cap Q) \,\Delta\, (P \cap R) &= (P \cap Q) \cup (P \cap R) \\
&\quad - (P \cap Q) \cap (P \cap R) \text{ (by definition)}
\end{aligned}
$$

Now:
$$
\begin{aligned}
(P \cap Q) \cup (P \cap R) &= P \cap (Q \cup R) && (D2) \\
\text{and} \quad (P \cap Q) \cap (P \cap R) &= (P \cap P) \cap (Q \cap R) && (C2) \\
&= P \cap (Q \cap R) && (A2) \\
\text{so} \qquad\qquad \text{RHS} &= [P \cap (Q \cup R)] - [P \cap (Q \cap R)] \\
&= P \cap [(Q \cup R) - (Q \cap R)] \text{ (Example 7)} \\
&= P \cap (Q \,\Delta\, R) \text{ (by definition)} \\
&= \text{LHS}
\end{aligned}
$$

Exercise 6d

Using the laws of the algebra of sets, prove:
1. $A - B = B' - A'$
2. $P \cup (Q - P) = P \cup Q$
3. $P - (Q \cap R) = (P - Q) \cup (P - R)$
4. $A \cup B \cup (A' \cap B') = \mathscr{E}$
5. $(P - Q) \cup (Q - P) = (P \cup Q) - (P \cap Q)$.

Exercises on Chapter 6

Write the dual of:
1. $P \cup (P' \cap Q) = P \cup Q$
2. $(X \cup \mathscr{E}) \cap (X \cap \emptyset) = \emptyset$
3. $(A \cap B) \cup (B \cap C) \cup (C \cap A) = (A \cup B) \cap (B \cup C) \cap (C \cup A)$
4. $(P \cap R) \cup (P' \cap Q) = (P \cup Q) \cap (P' \cup R)$

5. Draw a Venn diagram to illustrate the truth of the statement:

$$\text{If } P \cup Q = \mathscr{E}, \text{ then } P' \subseteq Q$$

   If instead $P \cap Q = \emptyset$, can the conclusion $P' \subset Q$ be drawn? If not, what is the true conclusion?
6. Write $A - B$ as an intersection. Draw a Venn diagram to show the relation $(A - B) \subset (A \cup B)$. What is the dual of this relation? Verify by a diagram that your answer is correct.

By a consideration of elements, prove:
7. $(A - B) \subset A$
8. If $P \cap Q = \emptyset$, then $Q \cap P' = Q$
9. $P - (P \cap Q) = P - Q$

Using the laws of the algebra of sets, prove:
10. $(P \cup Q)' \cup (P \cap Q) = (P' \cup Q) \cap (Q' \cup P)$
11. $[(P \cup Q)' \cup P]' = Q \cap P'$
12. $[X \cup (Y \cap Z)]' = X' \cap (Y' \cup Z')$
13. If $P \cap Q' = \emptyset$, then $P \triangle Q' = P \cup Q'$

14. Express in terms of intersection, union and complement:
    (a) $(P' \triangle Q)$ (b) $A - (B - C)$ (c) $P \cap (Q \triangle P)$

Exercises on Chapters 1–6

1. Let the shaded regions in Fig. 6.1 represent an operation in A, B denoted by $A*B$. Express $A*B$ in terms of union, intersection and complement. Prove:

   $A*B' = A'*B$

   $= (A \cap B') \cup (A' \cap B)$

   Is * a commutative operation?

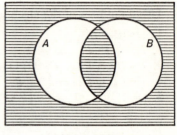

FIGURE 6.1

2. Prove $P - (Q - R) = P \cap (Q' \cup R)$
3. Prove $[P \cup (Q \cap R)]' = P' \cap (Q' \cup R')$
4. Prove $(P \triangle Q) = (P - Q) \cup (Q - P)$
5. Prove $(P - Q) - R = (P - R) - Q$
6. Prove $[P \cup (P \cup Q)']' = (P' \cap Q)$
7. Prove, using the laws of sets, that symmetric difference is a commutative operation.
8. Prove, using the laws of sets, that symmetric difference is an associative operation.
9. The product set $P \times P$ has nine elements, two being (x, y), (y, z). What are the other seven elements?
10. The *complement* of a composite set may be obtained by
    (*i*) replacing each constituent set by its complement
    (*ii*) interchanging $\cup$ and $\cap$
    (*iii*) interchanging $\mathscr{E}$ and $\emptyset$
    The complement of $A' \cup (B \cap C') \cup \emptyset$ is $A \cap (B' \cup C) \cap \mathscr{E}$.
    Write the complement of:
    (*a*) $(P \cup \emptyset) \cap (Q \cup P') \cap (P \cup Q' \cup \mathscr{E})$
    (*b*) $P \triangle Q$
    (*c*) $P \cup (Q \cap R) \cup (Q' \cap R') \cup (R \cap \emptyset)$
    (*d*) $(P - Q) \cup (P' \cap Q)$
11. Draw a carefully shaded Venn diagram to show whether each of the following statements is true or false:
    (*a*) intersection is right-distributive over difference
    (*b*) difference is left- and right-distributive over intersection
12. By finite induction, prove:

$$\frac{1}{1.4} + \frac{1}{4.7} + \frac{1}{7.10} + \cdots + \frac{1}{(3n-2)(3n+1)} = \frac{n}{3n+1}$$

13. Prove that the statement $(P \times Q) \times R = R \times (Q \times P)$ is false.
14. A lock is operated by rotating four rings each of which has eight different letters printed on its rim. When the word 'FIVE' appears, the lock opens. How many combinations will fail to open the lock?
15. If $\binom{2n}{4} = 21\binom{n}{2}$, find n.
16. Prove that, if:
$$P \subset R, \ Q \subset R, \ P \cap Q = \emptyset$$
then $\qquad (R - P) - Q = R - (P \cup Q)$
(This has already been verified by a Venn diagram in question 7 of the Exercises on Chapters 1 and 2, p. 23.)

Topics for investigation

Farey sequences

A Farey sequence (named after the man—a lawyer—who first investigated their properties) is the ordered set of rationals whose elements are:

(*i*) the zero rational, $\dfrac{0}{1}$

(*ii*) the irreducible proper fractions (arranged in order of magnitude) having denominators from 2 to n inclusive

(*iii*) the rational $\dfrac{1}{1}$

For example, $F_3 = \left\{\dfrac{0}{1}, \dfrac{1}{3}, \dfrac{1}{2}, \dfrac{2}{3}, \dfrac{1}{1}\right\}$. F_3 is said to be of order 3; the sequence F_n is said to be of order n. Satisfy yourself that there are no more rationals satisfying the requirements for F_3. Write out F_4 (which has 7 elements) and F_5 (which has 11). What are F_1 and F_2?

(*a*) If $\dfrac{p}{q}, \dfrac{r}{s}$ are consecutive terms of any given Farey sequence, investigate the value of $qr - ps$. Prove the result to be true for the general sequence F_n.

(*b*) If $\dfrac{p}{q}, \dfrac{x}{y}$ are two non-consecutive terms of a given Farey sequence F_n, investigate whether $\dfrac{p + x}{q + y}$ belongs to F_n, and if so under what condition. If $\dfrac{p+x}{q+y}$ does belong to F_n where is it placed

relative to $\dfrac{p}{q}$ and $\dfrac{x}{y}$?

(c) Investigate the truth of the statement $F_{n-1} \subset F_n$.

(d) If two elements t_r, t_{r+2} of F_n are given, can the element between them, t_{r+1}, be found, and if so, how? Can F_n be constructed completely if t_r and t_{r+2} are known?

The Pascal triangle

The first five rows of the 'triangle' (which takes its name from the French philosopher and mathematician, Blaise Pascal, 1623–62) are:

$$
\begin{array}{ccccccccc}
& & & & 1 & & & & \\
& & & 1 & & 1 & & & \\
& & 1 & & 2 & & 1 & & \\
& 1 & & 3 & & 3 & & 1 & \\
1 & & 4 & & 6 & & 4 & & 1 \\
\end{array}
$$

(a) What connection has the Pascal triangle with the binomial expansion?

(b) What are the next three rows of the triangle?

(c) Devise a rule for adding further rows to the triangle without making use of the binomial expansion.

(d) Show that the triangle illustrates the truth of the identity:

$$\binom{n}{r} = \binom{n-1}{r-1} + \binom{n-1}{r}$$

Triangular numbers

Let $S = \{1, 2, 3, 4, \ldots 16\}$. Write down the set S_1 of numbers obtained by adding the elements of S cumulatively, i.e., $\{1, 1+2, 1+2+3, \ldots, 1+2+3+\ldots+16\}$.

The elements of S_1 are called triangular numbers. Using dot patterns to represent the first few of these numbers, show why they are so called.

Write down the set S_0 whose elements are the odd elements of S. Write down the set S_2 of numbers obtained by adding the elements of S_0 cumulatively. What name would you give to the elements of S_2? Represent the first few of them by dot patterns.

Write down the set S_3 whose elements are the sums of all the consecutive pairs of elements of S_1. Illustrate by dot patterns what you observe about the sum of two consecutive triangular numbers. Write down the set S_E whose elements are the even elements of S. Write down the set S_4 of numbers obtained by adding the elements of S_E cumulatively. Devise a suitable name for the elements of S_4 and justify it by drawing dot patterns.

Write down the set S_5 obtained by removing from S every third element. Write down the set S_6 obtained by adding the elements of S_5 cumulatively.

Write down the set S_7 obtained by removing from S_6 every second element. Write down the set S_8 of numbers obtained by adding the elements of S_7 cumulatively. What name would you give to the elements of S_8? Represent the first two of them by dot patterns.

Suggest a method based on the above procedure by which a set of fourth powers could be obtained, and show that it works.

Does this method seem capable of generalization?

Cross-roads

Figure 6.2 represents a complex of roads running either N–S or E–W. Traffic on the N–S roads must travel S and traffic on the E–W roads must travel E. Copy the diagram, but without the lettering,

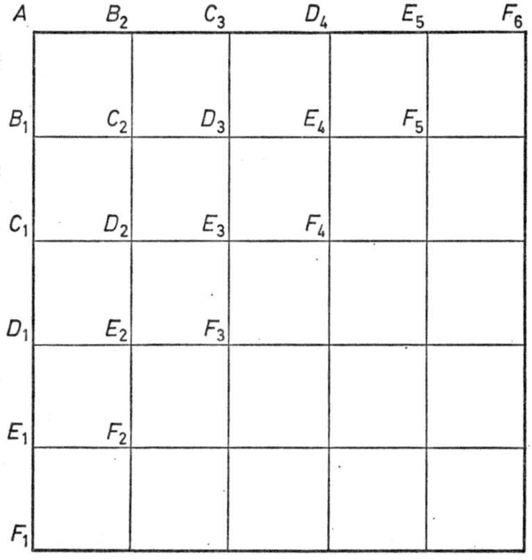

FIGURE 6.2

and in place of each letter fill in the number of ways of reaching that cross-roads from A. A itself will be replaced by 1 (why?).

Study the set of numbers you have obtained. Of what do they remind you? How is the connection to be explained?

Pythagorean triads

The numbers 3, 4, 5 are called a *Pythagorean triad* because a triangle whose sides are proportional to these three numbers is right-angled. Two more Pythagorean triads are 5, 12, 13 and 7, 24, 25. Study these three examples of such triads, and try to derive a method of obtaining as many more as may be desired.

7 | SYMBOLIC LOGIC

7.1 Sets of statements

Consider a group of five girls Alice, Beth, Connie, Doris and Elsie. About them a set of *statements* can be made: *e.g.*, 'Alice lives in Leeds, Beth lives in Leeds, Connie lives in Leeds, Doris lives in Leeds, Elsie lives in Leeds.' It may be that, in fact, Beth and Elsie live in Leeds, and the others do not: nevertheless, all five sentences are statements. Each has a *truth value*; it is either *true* or *false*.

On the other hand, the sentence 'She lives in Leeds' is not a statement, because the word 'she' is not defined, and therefore no truth value can be assigned to the sentence. For example, a ship is often referred to as 'she'; but the sentence 'The QE 2 lives in Leeds' is meaningless. 'She lives in Leeds' is an example of an *open sentence*. The distinction between an open sentence and a statement should be carefully noted; the open sentence does not have a truth value: the statement has.

Example 1. The following are examples of open sentences:

(*i*) It is no more than 20 km from London.
(*ii*) $x + 7 = 10$.
(*iii*) They make aircraft engines.

In (*i*) if 'it' is replaced by 'the brain' the sentence is meaningless; but the sentence becomes a statement when any place name is substituted for 'it'.

In (*ii*) if 'Edinburgh Castle' is substituted for x the sentence is meaningless; but $8 + 7 = 10$ is a statement, even though it is false.

In (*iii*) the open sentence becomes a statement if the name of a firm is put for 'they'.

Exercise 7a

Which of the following are open sentences, and which are statements? In the case of each statement, say whether it is true (T) or false (F):

(*a*) The moon reflects the light of the sun.
(*b*) It is 240,000 miles from the earth.
(*c*) $(x + 3)(x - 3) = x^2 - 9$.
(*d*) One hundred of them make one gramme.
(*e*) $1 + 1 + 1 + 1 = 0$.
(*f*) He could run a four-minute mile.
(*g*) It does over 20 km to the litre.
(*h*) Christmas Day is the last Sunday of December.
(*i*) Two of them were in the finals of the high jump.
(*j*) π is slightly greater than 3.

We are going to study statements in particular: they are often referred to as *propositions*.

Definition

A proposition is a sentence which is either true or false, but not both. This rules out questions (*e.g.*, 'How are you?') to which no truth value can be assigned.

It would be tedious if every written sentence were a single proposition, and so we use *connectives* to make the writing flow smoothly. Thus, 'Colin scored 24 runs. Nigel scored 8 runs.' becomes 'Colin scored 24 runs and Nigel, 8.' Each of these constituent statements is either true or false: but what of the compound proposition? We must examine this question for each of the connectives commonly used in English: we shall consider negation, conjuction, disjunction, condition and double condition.

7.2 Negation

Definition

Let p denote a simple proposition. The *negation* of p is 'not p', or an equivalent form.

Notation

We denote 'not p' by $\sim p$.

Negation does not imply more than one statement, and is therefore a *modifier* rather than a connective; but the distinction is unimportant. It is clear that if p is true, then $\sim p$ is false, and *vice versa*. The truth value of $\sim p$ may be exhibited in a *truth table* thus:

p	$\sim p$
T	F
F	T

A negation may be expressed in various forms of words, all equivalent.

Example 2. 'It is not true that daffodils bloom in spring' may be denoted by $\sim p$, where p denotes 'Daffodils bloom in spring'. 'Daffodils do not bloom in spring' or 'It is false that daffodils bloom in spring' would also be denoted by $\sim p$. What would $\sim (\sim p)$ denote?

7.3 Conjunction

Definition

A *conjunction* is a compound proposition formed from two or more simple propositions by the connective 'and', or an equivalent form.

Notation

Let p and q denote two simple propositions: then the conjunction of p and q is denoted by $p \wedge q$ (read as 'p and q').

We again define the truth value of a conjunction $p \wedge q$ in terms of the truth values of p and q by means of a truth table, thus:

p	q	$p \wedge q$
T	T	T
T	F	F
F	T	F
F	F	F

These are the only four logically possible combinations of truth values of p and q: the conjunction $p \wedge q$ is true if and only if p and q are both true.

In the above definition, the phrase 'or an equivalent form' should be noted. Think, for example, of the proposition 'April was a wet month, but May was drier'. It would be less common, but equally correct, to write 'and' instead of 'but': this is an example of a conjunction. So also is 'Roses are sometimes red: forget-me-nots, never' even though the connective 'and' is missing. What is the truth value of this last proposition?

Example 3. 'The battle of Waterloo, fought in 1816, was a decisive victory for the English.'

Here there are two constituent propositions. Let p be 'The battle of Waterloo was fought in 1816'; and let q be 'The battle of Waterloo was a decisive victory for the English.' Then the given statement is a conjunction, denoted by $p \wedge q$. p is false, q is true, hence $p \wedge q$ is false.

Example 4. $3^2 + 4^2 = 5^2$ and $5^2 + 12^2 = 13^2$. This compound proposition is true as both constituents are true.

Where there are three constituent propositions p, q, r, each of which may be true or false, there are eight possible combinations of truth values of p, q, r: they are TTT, TTF, TFT, TFF, FTT, FTF, FFT, FFF. Only in the first of these, where all three constituents are true, is the compound proposition true. Draw up a truth table for $p \wedge q \wedge r$.

Exercise 7b

State the truth value of each of the following propositions:
(a) $6 > -12$ and $-6 > -12$.
(b) 25 and 164 are perfect squares.
(c) London is the capital of England, and its largest city.
(d) The word *abstemious*, which means 'not given to indulgence', has all the vowels in alphabetical order.
(e) $1 \in N$ but $0 \notin N$.
(f) While birds have feathers, fish have scales.
(g) No prime number is even, and no perfect square is odd.
(h) Every square and every parallelogram is a rectangle.

7.4 Disjunction

Definition

A *disjunction* is a compound proposition formed from two or more simple propositions by the connective 'or'.

Notation

Let p and q denote two simple propositions: then the disjunction p or q is denoted by $p \vee q$ (read as 'p or q').

The definition requires amplification since the word 'or' has two possible meanings. If A says to B, 'I will phone you on Monday or Tuesday', he intends to make one phone call. But if a teacher asks, 'How many are taking physics or chemistry?' the number taking physics or chemistry or both is required. The first example illustrates the *exclusive* use of 'or' ('or' but not 'and'): the second illustrates the *inclusive* use of the word ('and/or'). In defining disjunction, therefore, it is necessary to state that the second use is intended:

$$p \vee q \text{ denotes } p \text{ and/or } q$$

Clearly a compound proposition in the form of a disjunction is true if one of the constituent statements is true or if both are. The truth table is:

p	q	$p \vee q$
T	T	T
T	F	T
F	T	T
F	F	F

The exclusive p or q (p or q but not p and q) is often denoted by $p \veebar q$. Symbolically:

$$p \veebar q = (p \vee q) \wedge \sim (p \wedge q)$$

Since $p \veebar q$ can be expressed in terms of $\sim$, $\wedge$, $\vee$ it is not necessary to include it among the essential basic connectives.

Write out the truth table for $p \veebar q$.

Example 5. Write symbolically: The committee may include X or Y, but not Z.

When assigning symbols to constituent statements for the purpose of expressing a compound proposition mathematically, it is usual to denote *positive* statements by $p, q, r \ldots$

Let p be 'The committee may include X'
  q be 'The committee may include Y'
and r be 'The committee may include Z.
Then the given proposition is denoted by $(p \lor q) \land {\sim}r$.

Example 6. If p stands for 'John takes PE', q for 'John takes music', r for 'John takes art', interpret:

(i) $p \land {\sim}(q \lor r)$ (ii) ${\sim}(p \lor q \lor r)$

Care should be taken to write the interpretation in idiomatic English:
 (i) might read, 'John takes PE, but not music or art'.
 (ii) might read 'John does not take PE or music or art'.

Exercise 7c

1. In each of the following compound propositions, denote positive constituent statements by $p, q, r, \ldots$ and express the proposition symbolically.
  (a) It is not true that $2 + 2 = 5$ and $3 + 3 = 6$.
  (b) It is neither sunny nor warm today.
  (c) Picasso painted this picture or I am a fool.
  (d) During this term the class will study two or three languages —French, and German or Spanish.
  (e) 15 is a multiple of 3, and 4 is a factor of 62, or 6 is not a prime number.
2. Let p be 'The house is warm'; q ,'The fuel supply is adequate'; r, 'The heating system is in good order'. Write in idiomatic English the propositions:
  (a) $q \land {\sim}r$ (d) ${\sim}q \lor (p \land r)$
  (b) $q \land r \land {\sim}p$ (e) ${\sim}p \land ({\sim}q \lor {\sim}r)$
  (c) $r \lor {\sim}p$

7.5 Condition

Definition

A *condition* is a compound proposition of the form 'if p then q' or an equivalent form.

Notation

'If *p* then *q*' is denoted by $p \rightarrow q$. The truth table for $p \rightarrow q$ is:

p	*q*	$p \rightarrow q$
T	*T*	*T*
T	*F*	*F*
F	*T*	*T*
F	*F*	*T*

It will be seen that a conditional compound proposition is false only if a false conclusion *q* is drawn from a true antecedent *p*.

For instance, if I say to someone: 'If you are good, I shall give you sweets', then I shall be false to my word only if he is good and I do not give him sweets.

Example 7

(*i*) 'If snow is white, then violets are blue' is a true proposition: the constituent statements have truth values *T*, *T*.

(*ii*) 'If 2 + 2 = 4, then − 6 > − 5' is false: the constituent statements have truth values *T*, *F*.

(*iii*) 'If M. Pompidou is president of the USA, then London is the capital of England' is true. The first statement is false and the second true.

(*iv*) 'If the dog is the largest known animal, then Tower Bridge is the longest bridge in the world.' This conditional proposition is true: its constituents have truth values *F*, *F*.

Again, it should be noted that various forms of words can be used to express condition.

Example 8. 'The coming of spring implies the return of the migrant birds.'

Let *p* be 'Spring comes'; and *q*, 'Migrant birds return'. Then the proposition is denoted by $p \rightarrow q$.

Example 9. 'French is more difficult than German if history is more interesting than geography.'

Let *p* be 'French is more difficult than German'; *q*, 'History is more interesting than geography'. Then the proposition is denoted by $q \rightarrow p$.

The fact that there is no necessary connection between the statements represented by p and q is irrelevant: in logic the one thing that matters is how the truth of the compound proposition depends on that of the constituent statements. Hence the truth table for each connective should be made the basis of all argument.

Exercise 7d

1. State the truth value of each of the following propositions:
   (a) If roses are red, violets are blue.
   (b) If $6 < -3$, then $-6 < 3$.
   (c) If all multiples of 7 are odd, then all multiples of 9 are even.
   (d) Paris is the capital of Sweden if Madrid is the capital of Spain.
   (e) That there are 24 hours in a day implies that there are 100p in £1.
2. If p represents 'I work hard'; q, 'I succeed in my career'; r, 'I take sufficient recreation'; interpret the propositions:
   (a) $\sim p \to \sim q$ (d) $\sim r \wedge q$
   (b) $(p \wedge \sim r) \to \sim q$ (e) $\sim (r \vee p)$
   (c) $q \to (p \wedge r)$
3. Express in terms of p, q, r (question 2) the propositions:
   (a) It is not true that I work hard and succeed in my career.
   (b) I do not take sufficient recreation nor do I work hard, and I succeed in my career.
   (c) I succeed in my career if I take sufficient recreation.
   (d) I work hard and succeed in my career if I do not take sufficient recreation.
   (e) If I take sufficient recreation and do not succeed in my career, then I do not work hard.

7.6 Double condition

In a number of places in this book the phrase 'if and only if' has occurred. This phrase must now be examined.

Definition

A *double condition* is a compound proposition of the form 'p if and only if q'.

Notation

'*p* if and only if *q*' is denoted by $p \leftrightarrow q$. The phrase 'if and only if' is often contracted to 'iff'.

The distinction between the simple condition 'if' and the double condition 'iff' may be brought out by an example.

Example 10. Consider the statements:

   (*i*) We shall cancel the picnic if it rains.
   (*ii*) We shall cancel the picnic only if it rains.
   (*iii*) We shall cancel the picnic if and only if it rains.

Let *p* be 'We shall cancel the picnic'; *q*, 'It rains'.

   (*i*) says that rain is a *sufficient condition* for cancellation. Even if there is no rain the picnic may still be cancelled for some other reason; but the statement takes no account of this: it says simply, 'if *q* then *p*' *i.e.*, $q \rightarrow p$.

   (*ii*) says that no other condition for cancellation will be considered: rain is a *necessary condition*. If, therefore, the picnic is cancelled, then it must be raining: *i.e.*, $p \rightarrow q$.

   (*iii*) says both of these: rain is a *necessary and sufficient condition* for cancellation. Symbolically, the statement is, 'if *p* then *q*, and if *q* then *p*' or $(p \rightarrow q) \wedge (q \rightarrow p)$.

This double condition is written for short $p \leftrightarrow q$.

The truth table for $p \leftrightarrow q$ is:

p	q	$p \leftrightarrow q$
T	*T*	*T*
T	*F*	*F*
F	*T*	*F*
F	*F*	*T*

Note that since $p \leftrightarrow q$ is equivalent to a conjunction the *two* conclusions $p \rightarrow q$ and $q \rightarrow p$ follow from it.

Example 11. Express symbolically: 'Don will bowl and Bill will bat only if the state of the game requires it.'

Let *p* be 'Don will bowl'; *q*, 'Bill will bat'; *r*, 'The state of the game requires it'. Then the compound statement is denoted by $(p \wedge q) \rightarrow r$.

D

Example 12. Express symbolically: 'Students who take chemistry or biology will be present at the lecture if they have no other class at that time.'

Let p be 'A student takes chemistry'; q, 'A student takes biology'; r, 'A student attends the lecture'; s, 'A student has no other class at that time.' Then the compound statement may be written:

$$[(p \vee q) \wedge s] \to r$$

Example 13. Consider the statement of Section 7.3 (p. 86): 'The conjunction $(p \wedge q)$ is true if and only if both p and q are true.' That is:

$$(p \wedge q \text{ is true}) \leftrightarrow [(p \text{ is true}) \wedge (q \text{ is true})]$$

This leads to the double conclusion:

(*i*) $(p \wedge q \text{ is true}) \to [(p \text{ is true}) \wedge (q \text{ is true})]$
(*ii*) $[(p \text{ is true}) \wedge (q \text{ is true})] \to (p \wedge q \text{ is true})$

7.7 The sufficiency of these connectives

We have been considering compound propositions formed by the use of various connectives. There have been five of these (including negation which is commonly classed with the connectives). The question arises: Do negation, conjunction, disjunction, condition and double condition cover all the possibilities? In other words, might there be a compound proposition which could not be stated symbolically in terms of the connectives $\sim$, $\wedge$, $\vee$, $\to$, $\leftrightarrow$?

Let there be two constituent statements p, q; then, as we have seen, there are four possible combinations of truth value: *TT, TF, FT, FF*. When p, q are combined by one of the connectives, the compound proposition so formed will be true or false in each of these four cases: there are therefore sixteen possible sets of entries in the final column of the truth table. They are:

(*a*)	(*b*)	(*c*)	(*d*)	(*e*)	(*f*)	(*g*)	(*h*)
TTTT,	*TTTF,*	*TTFT,*	*TTFF*	*TFTT,*	*TFTF,*	*TFFT,*	*TFFF,*
FFFF,	*FFFT,*	*FFTF,*	*FFTT,*	*FTFF,*	*FTFT,*	*FTTF,*	*FTTT.*

The last eight of these are the negations of the first eight (why?): we have therefore only to consider the eight cases on the top line.

Are all these covered by our set of connectives? With the exception of $TTTT$, it is easy to see that they are. Study the following table which demonstrates the fact:

(d)	(f)	(h)	(b)	(e)	(g)	(c)
p	q	$p \wedge q$	$p \vee q$	$p \rightarrow q$	$p \leftrightarrow q$	$q \rightarrow p$
T	T	T	T	T	T	T
T	F	F	T	F	F	T
F	T	F	T	T	F	F
F	F	F	F	T	T	T

Then, on the lower line, $FFFT$ is the set of truth values of $\sim(p \vee q)$, and so on. Draw up a similar table for this line.

The special cases of the truth tables whose final column reads $TTTT$ or $FFFF$ will be considered in the next chapter.

Exercises on Chapter 7

1. Prove that a truth table involving n simple propositions has 2^n rows.
2. (a) If p is a necessary condition for q, is q a necessary condition for p?
   (b) If p is a sufficient condition for q, is q a necessary condition for p?
   (c) If p is a necessary and sufficient condition for q, is q a necessary and sufficient condition for p?
   (d) If p is a necessary condition for q, is q a sufficient condition for p?
   (e) If p is a sufficient condition for q, is q a sufficient condition for p?
3. Make (i) a true statement, (ii) a false statement, from each of the following open sentences:
   (a) $x^2 - 3x = 4$.
   (b) It is a piece of land entirely surrounded by water.
   (c) Not only did they fail to come, but they sent no explanation.
   (d) Some x are y.
   (e) A is in the same Form as B.

4. If p and q denote two statements, the first true and the second false, write down the truth values of:

(a) $\sim(p \lor q)$ (e) $(p \land q) \lor \sim p$

(b) $\sim p \land \sim q$ (f) $\sim q \lor (p \to q)$

(c) $(p \to \sim q)$ (g) $(\sim p \to \sim q) \lor (p \land q)$

(d) $(\sim p \leftrightarrow \sim q)$

5. In each of the following propositions, state whether the disjunction is inclusive or exclusive:

(a) I scored 71 or 72 in my class test.

(b) A prize is offered for good work or regular attendance.

(c) John will take exercise or he will become ill.

(d) Heather will be in her study or in the library this morning.

6. State the truth value of each of the following propositions:

(a) $(6 > -8) \to (-6 < 8)$

(b) $[(x^2 - 1) = (x + 1)^2 - 2(x + 1)]$ iff $[x + 1 = x - 1]$ $(x \in R)$

(c) $(x \in N)$ only if $(x \in Q)$

(d) $(Z \subset Q) \lor (R \subset Q)$

7. Denote by p, q, r the positive simple statements from which the following propositions are derived, and write the propositions symbolically:

(a) It is not true that if the barometer is falling it will rain.

(b) The train will not be on time if and only if the signals are against it.

(c) Poor visibility is a sufficient condition for road accidents.

(d) If I am not to be discouraged, it is necessary that I be given due credit.

8. Let p be 'The zoo is well stocked with animals'; q, 'Entrance charges are high'; r, 'Attendances are large'. Write in idiomatic English:

(a) $p \to (q \land r)$ (c) $r \leftrightarrow (p \lor \sim q)$

(b) $(\sim r \lor \sim q) \to p$ (d) $(\sim p \land q) \to \sim r$

9. If $p \to q$ has truth value T and $p \land q$ has truth value F, what is the truth value of p?

10. If $p \lor q$ has truth value T and $q \to \sim p$ has truth value F, what is the truth value of $p \leftrightarrow q$?

11. If $p \leftrightarrow q$ has truth value T and $\sim(p \land q)$ has truth value T, what is the truth value of $p \lor q$?

12. Refer to the following Sections where the phrase 'if and only if' has been used, and analyse its significance as a *double* condition in each case:

(a) Section 1.3, p. 4   (c) Section 4.1, p. 45

(b) Section 3.3, p. 37   (d) Section 7.3, p. 86

8 | COMPOUND PROPOSITIONS

8.1 Compound propositions

When a statement involves a single connective its truth value is given, as we have seen, by a truth table. When more than one connective is required to express the proposition symbolically, the truth table for the compound proposition is drawn up by combining the truth tables for the different connectives involved.

Example 1. Draw up a truth table for the proposition $p \to (p \wedge q)$.
The table may take one of two forms:

p	q	$p \wedge q$	$p \to (p \wedge q)$
T	T	T	T
T	F	F	F
F	T	F	T
F	F	F	T
(1)	(2)	(3)	(4)

Column (3) is a basic truth table. Column (4) follows by applying the truth table for $\to$ to columns (1) and (3). Column (4) is then the truth table of the given proposition.

This arrangement has the advantage that the table required appears in the final column.

In the second arrangement, which is shown in the table at the top of the next page, the proposition is set out in the form in which it is given: columns (3), (4), (5) are completed in that order.

This arrangement does not show the required truth values in the final column, and for that reason the column which provides the answer may be indicated by *.

p	q		p	$\rightarrow$	$(p \wedge q)$
T	T		T	T	T
T	F		T	F	F
F	T		F	T	F
F	F		F	T	F
(1)	(2)		(3)	(5)*	(4)

This second form of table is perhaps preferable to the first as being more direct: but it is wise to begin by numbering the columns in the order in which they will be completed.

Example 2. The truth table of the proposition $p \rightarrow [(p \wedge q) \vee (q \wedge r)]$ is:

p	q	r		p	$\rightarrow$	$[(p \wedge q)$	$\vee$	$(q \wedge r)]$
T	T	T		T	T	T	T	T
T	T	F		T	T	T	T	F
T	F	T		T	F	F	F	F
T	F	F		T	F	F	F	F
F	T	T		F	T	F	T	T
F	T	F		F	T	F	F	F
F	F	T		F	T	F	F	F
F	F	F		F	T	F	F	F
(1)	(2)	(3)		(4)	(8)*	(5)	(7)	(6)

Column (8) shows the compound proposition to be true except when p is true and q is false.

Definition

Two compound propositions having the same constituent statements p, q, r . . . are *equivalent* if and only if they have the same truth values for all the possible combinations of truth values of p, q, r, . . .

Notation

If P_1 and P_2 are equivalent propositions, we write $P_1 = P_2$.

Example 3. Prove that:

$$[p \lor (q \land r)] \lor [(q \land \sim p) \lor p] = p \lor [q \land (\sim p \lor r)]$$

The truth tables are:

p	q	r		$[p$	$\lor$	$(q \land r)]$	$\lor$	$[(q \land \sim p)$	$\lor$	$p]$
T	T	T		T	T	T	T	F	T	T
T	T	F		T	T	F	T	F	T	T
T	F	T		T	T	F	T	F	T	T
T	F	F		T	T	F	T	F	T	T
F	T	T		F	T	T	T	T	T	F
F	T	F		F	F	F	T	T	T	F
F	F	T		F	F	F	F	F	F	F
F	F	F		F	F	F	F	F	F	F
(1)	(2)	(3)		(4)	(6)	(5)	(10)	(7)	(9)	(8)
							*			

	p	q	r		p	$\lor$	$[q$	$\land$	$(\sim p \lor r)]$
	T	T	T		T	T	T	T	T
	T	T	F		T	T	T	F	F
	T	F	T		T	T	F	F	T
	T	F	F		T	T	F	F	F
	F	T	T		F	T	T	T	T
	F	T	F		F	T	T	T	T
	F	F	T		F	F	F	F	T
	F	F	F		F	F	F	F	T
	(1)	(2)	(3)		(11)	(15)	(12)	(14)	(13)
						*			

Column (10) is derived from columns (6) and (9): column (15) is derived from columns (11) and (14). The fact that columns (10) and (15) are identical proves that the propositions are equivalent.

It is, of course, usual to combine the tables, thus avoiding the repetition of columns (1), (2), and (3). With practice, columns (4), (8), and (11) which are repetitions of (1) may also be omitted.

Exercise 8a

Draw up the truth tables of the following propositions:

1. $(p \land \sim q) \to q$
2. $(p \lor q) \leftrightarrow (\sim p \land q)$

3. $\sim[p \to (q \to p)]$
4. $(p \wedge \sim q) \to (r \wedge \sim r)$
5. $p \to [p \wedge \sim(q \vee r)]$

8.2 Tautology and contradiction

Example 4. Consider the proposition $(\sim p \wedge \sim q) \to [p \to (\sim p \vee q)]$.
The truth table is:

p	q	$(\sim p \wedge \sim q)$	$\to$	$[p$	$\to$	$(\sim p \vee q)]$
T	T	F	T	T	T	T
T	F	F	T	T	F	F
F	T	F	T	F	T	T
F	F	T	T	F	T	T
(1)	(2)	(3)	(7)*	(4)	(6)	(5)

Observe that the proposition is true in all cases.

Definition

A proposition which is true for all the possible combinations of truth values of its constituent statements is called a *tautology*.

Notation

A tautology will be denoted by t.

Example 5. Examine the truth value of the proposition:
$$(p \wedge q) \leftrightarrow (\sim p \vee \sim q)$$
The truth table is:

p	q	$(p \wedge q)$	$\leftrightarrow$	$(\sim p \vee \sim q)$
T	T	T	F	F
T	F	F	F	T
F	T	F	F	T
F	F	F	F	T
(1)	(2)	(3)	(5)*	(4)

It is seen that this proposition is false in all cases.

Definition

A proposition which is false for all the possible combinations of truth values of its constituent statements is called a *contradiction*.

Notation

A contradiction will be denoted by f.

Note that tautology and contradiction are the two special cases referred to in Section 7.7, p. 93.

Exercise 8b

Prove:

1. $[p \wedge (q \vee \sim q)] \leftrightarrow p = t$
2. $(p \to q) \leftrightarrow (p \wedge \sim q) = f$
3. $[(p \to q) \to q] \to (p \vee q) = t$
4. $[p \wedge (p \to q)] \wedge \sim q = f$
5. $\sim (p \to q) \to [(q \vee r) \to (p \vee r)] = t$

8.3   The truth set of a proposition

Consider the statement 'Joan is a keen hockey player', and imagine all the girls of that name brought together. This is the universal set of girls named Joan. It can be partitioned into two sets consisting in one case of those who are keen hockey players, and in the other, of those who are not. The set of girls for whom the statement is true is called the *truth set* of the statement; and if we denote the statement by p, its truth set may be denoted by P.

The other set of Joans for whom the statement is false is clearly the complement of P, *i.e.*, P': P' is the truth set of the statement $\sim p$.

To every statement p there can similarly be assigned a set P, P being the truth set of p.

Now consider the compound proposition 'The child is both musical and artistic'. Here the universal set is the set of all children. If p denotes 'The child is musical' and q, 'The child is artistic', then the proposition 'The child is both musical and artistic' is denoted by $p \wedge q$. Let P, Q be the truth sets of p, q respectively. It is clear that the truth set of $p \wedge q$ will have as its elements only those children who satisfy both conditions of being musical and being artistic: *i.e.*, the truth set of $p \wedge q$ is $P \cap Q$.

In the same way, the truth set of $p \lor q$ is $P \cup Q$. Think of an example to illustrate this.

The other two common connectives, $\rightarrow$ and $\leftrightarrow$ can both be expressed in terms of the three we have already considered.

Exercise 8c

1. Draw up the truth table of the proposition $\sim p \lor q$, and compare it with that of $p \rightarrow q$. What conclusion do you draw? If P, Q are the truth sets of p, q, respectively, write down the truth set of $p \rightarrow q$.
2. Draw up the truth table of the proposition:
$$(p \land q) \lor (\sim p \land \sim q)$$
and compare it with that of $p \leftrightarrow q$. What conclusion do you draw? If P, Q are the truth sets of p, q, respectively, write down the truth set of $p \leftrightarrow q$.

We may thus summarize our results:

Statement	Truth Set
p	P
q	Q
$p \land q$	$P \cap Q$
$p \lor q$	$P \cup Q$
$p \rightarrow q$	$P' \cup Q$
$p \leftrightarrow q$	$(P \cap Q) \cup (P' \cap Q')$

8.4 Propositions and Venn diagrams

The fact that to any proposition there corresponds a set suggests that Venn diagrams may be of value in examining compound statements.

Example 6. Write down the truth table of the statement $p \rightarrow q$. Its truth set, $P' \cup Q$, is shown in Fig. 8.1, being represented by the shaded regions. Check this.

Observe that only the unshaded area represents the case where p is true and q is false, and this is the one case in which $p \rightarrow q$ is false. The area shaded obliquely represents the case where both p and q are true and $p \rightarrow q$ is true. What cases do the regions shaded vertically and horizontally represent?

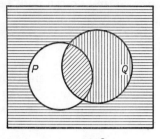

FIGURE 8.1   FIGURE 8.2

Exercise 8d

Draw shaded Venn diagrams showing the truth sets of the statements:

(a) $p \wedge q$ (b) $p \vee q$ (c) $p \leftrightarrow q$

and examine them in the light of the truth tables of these connectives.

Example 7. Draw a shaded Venn diagram to show the truth set of the proposition $(p \wedge q) \rightarrow (p \vee q)$.

We first write $(p \wedge q) \rightarrow (p \vee q)$ in the form $\sim(p \wedge q) \vee (p \vee q)$: then the truth set is $(P \cap Q)' \cup (P \cup Q)$. This is illustrated in Fig. 8.2 where $(P \cap Q)'$ is shaded vertically, and $(P \cup Q)$ horizontally.

It is seen that the union of these two sets is $\mathscr{E}$, since there is no part of the rectangle unshaded. Hence there is no case in which the proposition is false, and it is therefore a tautology. Verify this by means of a truth table.

Exercise 8e

By drawing carefully shaded Venn diagrams, decide which, if any, of the following propositions are tautologies:

1. $p \rightarrow (\sim p \rightarrow q)$
2. $\sim(p \vee q) \wedge p$
3. $\sim p \rightarrow (p \rightarrow q)$
4. $[p \wedge (p \rightarrow q)] \rightarrow q$
5. $(p \rightarrow q) \leftrightarrow \sim(p \wedge \sim q)$

8.5 Set theorems and truth tables

In Chapter 6 two methods of proving set theorems were studied, one by a consideration of elements, and the other by means of the laws of set algebra. The connection between propositions and set theory which has just been noticed leads to a third method of proving theorems on sets—by the use of truth tables.

Example 8. Prove DeMorgan's Laws (E5, E6, p. 72).
If we prove $(P \cap Q)' = P' \cup Q'$, the dual will follow.
Let p, q be statements whose truth sets are P, Q, respectively: then the statement whose truth set is $(P \cap Q)'$ is $\sim(p \wedge q)$: and the statement whose truth set is $P' \cup Q'$ is $\sim p \vee \sim q$. The given sets will be equal if the corresponding propositions are equivalent. We therefore prove that $\sim(p \wedge q)$ and $\sim p \vee \sim q$ are equivalent, *i.e.*, that they have the same truth table.
The truth values of $\sim(p \wedge q)$ are *FTTT* (why?). The truth table of $\sim p \vee \sim q$ is:

p	q	$\sim p$	$\vee$	$\sim q$
T	T	F	F	F
T	F	F	T	T
F	T	T	T	F
F	F	T	T	T
(1)	(2)	(3)	(5)	(4)
			*	

Hence the law is proved.

Exercise 8f

Prove the following set theorems by means of truth tables:
1. $P \cup Q = (P \cap Q') \cup Q$
2. $P \cup (Q \cap P) = P$
3. $(P \cup Q) \cap R' = [(P \cup Q)' \cup R]'$
4. $P \cup (Q \cap R) = (P \cup Q) \cap (P \cup R)$
5. $(P \cap Q \cap R) \cup (P' \cup Q' \cup R') = \mathscr{E}$

Exercises on Chapter 8

1. Draw up the truth table of the proposition:
$$(p \lor \sim q) \to \sim (q \lor \sim p).$$
What is the truth value of this proposition:
   (a) when p is true and q is false
   (b) when both p and q are false

2. Prove by means of a truth table that:
$$[p \lor (\sim p \lor q)] \lor (\sim p \land \sim q)$$
is a tautology. What can you say about:
$$\sim \{[p \lor (\sim p \lor q)] \lor (\sim p \land \sim q)\}$$

3. If the truth sets of statements p, q, are P, Q, respectively, write the truth set of the proposition $(\sim p \lor \sim q) \to (p \land \sim q)$. Draw a Venn diagram, shading the regions which represent this truth set. From the diagram state the truth value of the proposition when p is false.

4. By means of a truth table prove law D2 (p. 72) of set algebra.

5. If P, Q are the truth sets of statements p, q, respectively, draw a Venn diagram shading the regions representing the truth sets of $p \leftrightarrow q$ and $\sim p \lor q$. Deduce that $(p \leftrightarrow q) \to (p \to q)$ is a tautology.

6. Prove by any valid method that the statements $(\sim p) \leftrightarrow q$ and $(p \leftrightarrow q) \to (\sim p \land q)$ are equivalent.

7. Prove that $\sim (p \to q) = (p \land \sim q)$. Illustrate the equivalence by Venn diagrams.

8. Prove that $p \veebar q$ (Section 7.4, p. 87) is equivalent to:
$$(p \lor q) \land \sim (p \land q)$$

9. Prove that $(p \veebar q) \veebar (p \land q) = p \lor q$.

10. What is the truth value of the proposition:
$$[(p \to r) \land (q \to r)] \to (p \leftrightarrow q)$$
   (a) when p and q are true and r is false
   (b) when p is true and q and r are false

Exercises on Chapters 7 and 8

1. Draw up the truth table of the proposition:
$$(p \land \sim q) \leftrightarrow \sim (q \to \sim p)$$
What is the truth value of this proposition:
   (a) when p is true and q is false
   (b) when both p and q are false

2. If the proposition $\sim(p \wedge q) \to (p \vee q)$ is true, deduce the truth value of each of the propositions:

(a) $(p \wedge q) \vee (p \vee q)$

(b) $\sim(p \wedge q) \wedge \sim(p \vee q)$

(c) $(p \wedge q) \vee [(p \wedge q) \to (p \vee q)]$

3. Fig. 8.3 shows (shaded) the truth set of a compound proposition involving statements p, q, r.

(a) Write down the proposition.

(b) Draw up its truth table.

(c) Identify each of the eight regions of the Venn diagram with a row of the truth table. To which region does the second row of the table correspond?

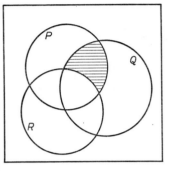

FIGURE 8.3

4. Is p a sufficient condition for $p \vee q$? (Hint: draw up the truth table of $p \to (p \vee q)$.) Is p a necessary condition for $p \vee q$? Draw Venn diagrams of the corresponding truth sets to illustrate your answers.

5. Prove that the proposition $[(\sim p \to q) \vee (r \vee \sim p)] \to (q \to r)$ is false when, and only when, q is true and r is false.

6. If the proposition $(p \wedge q) \to [(p \vee q) \to r]$ is false, prove that p and q are true and r is false. Is it sufficient to prove that when p and q are true and r is false, the proposition is false?

7. Prove, by any valid method, $[(p \veebar \sim q) \vee (\sim p \veebar q)] = p \leftrightarrow q$.

8. Draw up the truth table of $[\sim p \to (q \vee r)] \vee [p \to (\sim q \vee r)]$. What conclusion do you draw from the table?

9. Let $p \to \sim q$ be denoted by $p \downarrow q$.

(i) Draw up the truth table of $p \downarrow q$.

(ii) Prove $p \vee q = \sim p \downarrow \sim q$.

(iii) Express in terms of $\downarrow$:

    (a) $p \to q$ (c) $q \to \sim p$

    (b) $p \vee \sim q$ (d) $\sim(p \wedge q)$

Is the operation denoted by $\downarrow$ a commutative operation?

10. Is the operation denoted by $\downarrow$ (see question 9) an associative operation? [Ask: Is $p \downarrow (q \downarrow r)$ equivalent to $(p \downarrow q) \downarrow r$?]

11. With the same definition of $\downarrow$, test whether $\downarrow$ distributes from the left over $\wedge$. [Ask: Is $p \downarrow (q \wedge r)$ equivalent to $(p \downarrow q) \wedge (p \downarrow r)$?]

12. With the same definition of $\downarrow$, test whether $\downarrow$ is left-distributive over $\vee$.

13. Prove that the proposition $[\sim p \vee (r \vee \sim q)] \wedge [p \vee (r \vee q)]$ is true except when p and q are both true or both false and r is false.

14. Using a truth table, prove that the set $(P \cap Q') \cap (P' \cap Q)$ is empty.

15. Prove in three distinct ways that $(\sim p) \leftrightarrow q = \sim(p \leftrightarrow q)$.

16. Prove that the operation denoted by $\veebar$ is both commutative and associative.

17. Prove that the truth set of the proposition $p \leftrightarrow q$ may be written in the form $(P' \cup Q) \cap (P \cup Q')$.

9 | VALID ARGUMENT

9.1 Valid argument

An *argument* consists of a set of statements leading from an initial proposition (or a set of initial propositions), assumed to be true, to a final proposition.

The initial propositions on which the argument is based are called the *premises*: the final proposition is the *conclusion*. An argument may be *valid* or *invalid* (*i.e., fallacious*).

An argument is valid if and only if the statements comprising it follow logically one upon another. The proving of geometrical propositions, in particular, depends upon valid argument in which each statement (or *deduction*) is justified by a premise (or *hypothesis*), or by an axiom, or a theorem already proved true, or by a valid deduction already made.

Note that the validity of an argument in no way depends upon the truth or falsity of the conclusion. For example, it would be possible, by an appropriate choice of premises, to produce a valid argument that ice is heavier than water. The test of validity is simply whether the conclusion follows logically from the premises.

Example 1. Does $\sim q \to \sim p$ follow logically from $p \to q$? In other words, if $p \to q$, is it a valid argument to say: 'Therefore $\sim q \to \sim p$'?

We have to examine the compound statement 'If $p \to q$, then $\sim q \to \sim p$'; or, symbolically:

$$(p \to q) \to (\sim q \to \sim p)$$

The argument will be valid if and only if this proposition is a tautology. We can test this by drawing up a truth table:

E   107

p	q	$(p \to q)$	$\to$	$(\sim q$	$\to$	$\sim p)$
T	T	T	T	F	T	F
T	F	F	T	T	F	F
F	T	T	T	F	T	T
F	F	T	T	T	T	T
(1)	(2)	(3)	(7)	(4)	(6)	(5)

The proposition is therefore always true, and the argument is valid.

In this example we may go further, for the truth values of $p \to q$ (column 3) are the same as those of $(\sim q \to \sim p)$ (column 6): hence the two statements are equivalent, and we may write:

$$(p \to q) = (\sim q \to \sim p)$$

Definition

The statement $\sim q \to \sim p$ is called the *contrapositive* of $p \to q$. A statement and its contrapositive are equivalent.

Exercise 9a

Write the contrapositive of each of the following statements:

(a) $p \to \sim q$ (d) $(p \wedge q) \to r$
(b) $\sim p \to q$ (e) $(p \vee q) \to \sim(q \vee r)$
(c) $\sim p \to \sim q$ (f) $\sim p \vee q$

Example 2. Test the validity of the argument: 'If $p \to q$, then $q \to p$.' We ask 'Is $(p \to q) \to (q \to p)$ a tautology?' The truth table is:

p	q	$(p \to q)$	$\to$	$(q \to p)$
T	T	T	T	T
T	F	F	T	T
F	T	T	F	F
F	F	T	T	T
(1)	(2)	(3)	(5)	(4)

Column 5 shows that the proposition is not a tautology, hence the argument is invalid: $(q \to p)$ does not follow logically from $(p \to q)$.

Definition

The statement $q \to p$ is called the *converse* of $p \to q$. It is not valid to argue from a premise to its converse.

Example 3. Examine the validity of the argument:

$$p \to q$$
$$r \to \sim q$$
$$\overline{}$$
$$p \to \sim r$$

(The two premises are written above the line and the conclusion is written below.)

The argument is:

'If $(p \to q)$ and $(r \to \sim q)$, then $(p \to \sim r)$'
or $\quad [(p \to q) \land (r \to \sim q)] \to (p \to \sim r)$

The truth table is:

p	q	r	$[(p \to q)$	$\land$	$(r \to \sim q)]$	$\to$	$(p \to \sim r)$
T	T	T	T	F	F	T	F
T	T	F	T	T	T	T	T
T	F	T	F	F	T	T	F
T	F	F	F	F	T	T	T
F	T	T	T	F	F	T	T
F	T	F	T	T	T	T	T
F	F	T	T	T	T	T	T
F	F	F	T	T	T	T	T
(1)	(2)	(3)	(4)	(6)	(5)	(8)	(7)
						*	

Verify the table. Is the argument valid?

Note that in this case columns (6) and (7) differ, *i.e.*:

$$[(p \to q) \land (r \to \sim q)] \text{ and } (p \to \sim r)$$

are not equivalent, but the second statement does follow logically from the first.

Exercise 9b

Show by means of truth tables that each of the following arguments is valid:

(a) p
   $\dfrac{p \to q}{q}$

(c) $p \to q$
   $\dfrac{\sim q}{\sim p}$

(e) $p \to q$
   $p \lor r$
   $\dfrac{\sim r}{q}$

(b) $p \to q$
   $\dfrac{q \to r}{p \to r}$

(d) $\sim p \lor q$
   $\dfrac{\sim r \lor \sim q}{\sim p \lor \sim r}$

9.2   Two common valid arguments

Exercises (a) and (b) above introduce two types of valid argument which are often useful in testing the validity of other arguments.

(A) p
        $\dfrac{p \to q}{q}$ is called the law of *detachment*

(B) $p \to q$
        $\dfrac{q \to r}{p \to r}$ is called the law of *syllogism*

Along with these should be remembered two equivalences already met with:

(C) $p \to q \;=\; \sim q \to \sim p$ (the contrapositive)
(D) $p \to q \;=\; \sim p \lor q$

Example 4.   Test the validity of the argument:

$$\sim p \to \sim q$$
$$\dfrac{p \to r}{\sim r \to \sim q}$$

In this case it is possible to avoid drawing up a truth table, for:

$\qquad\qquad \sim p \to \sim q = q \to p$ (contrapositive)
then $\qquad q \to p$ and $p \to r \Rightarrow q \to r$ (syllogism)
$\qquad\qquad\qquad\qquad\quad \Rightarrow \sim r \to \sim q$ (contrapositive)

Even where it is found necessary to draw up a truth table, it may be possible to simplify the argument first, thus reducing the tabular work involved.

Example 5. Test the validity of the argument:

$$p \to q \qquad (i)$$
$$(p \to r) \to s \qquad (ii)$$
$$\underline{\sim s \qquad \qquad} (iii)$$
$$q$$

(*ii*) may be replaced by its contrapositive $\sim s \to \sim(p \to r)$ and this
with $\sim s$ gives $\sim(p \to r)$ (detachment). The argument thus reduces to:

$$p \to q$$
$$\underline{\sim(p \to r)}$$
$$q$$

Show by means of a truth table that this argument is valid.

Note that the original argument involves four basic statements
p, q, r, s; and, had a truth table been drawn up immediately, it would
have had 16 rows. Why?

Exercise 9c

1. Express each of the following statements as a condition:
   (a) $\sim p \lor \sim q$ (c) $p \lor q$
   (b) $p \lor \sim q$
2. Express each of the following statements as a disjunction:
   (a) $p \to \sim q$ (c) the contrapositive of $\sim p \to \sim q$
   (b) $\sim p \to q$ (d) the converse of $\sim q \to \sim p$
3. Test the validity of the following arguments, making as much
   use as possible of A, B, C, D, Section 9.2:
   (a) $p \to \sim q$ (d) $\sim q \to \sim p$

         q $q \to \sim r$

      $\underline{\sim r \to p}$ $\underline{p \to \sim r}$

         r

   (b) p (e) $\sim p$

      $p \to q$ $r \to p$

   $\underline{\sim q \lor \sim r}$ $\underline{\sim r \to \sim q}$

        $\sim r$ $\sim q$

   (c) p

         q

   $\underline{r \to \sim(p \land q)}$

        $\sim r$

9.3 Indirect proof

If instead of proving a given proposition to be true we prove an equivalent proposition true, we have an *indirect* proof of the given proposition. An argument may be proved valid indirectly in several ways.

First method. We may prove the contrapositive of the given proposition: then that proposition will follow.

Example 6. Prove that the following argument is valid: If n^2 is an even integer, then n is an even integer.

Instead we may prove that if n is *not* an even integer then n^2 is not an even integer. But this is so, for, if $n = 2m + 1$ (an odd integer), then $n^2 = (4m^2 + 4m) + 1$ (also an odd integer).

Example 7. Establish the validity of the argument:

$$p \to r \quad (i)$$
$$\underline{q \to \sim r} \quad (ii)$$
$$q \to \sim p$$

From (ii), $r \to \sim q$ (contrapositive); then $p \to r$ and $r \to \sim q$ give $(p \to \sim q)$ (syllogism).

We have proved the contrapositive of the given conclusion, whence that conclusion follows.

Second method. We may assume the conclusion to be false, and show that this assumption contradicts one of the premises.

Example 8. Test the validity of the argument:

$$p \vee q$$
$$\underline{\sim q}$$
$$p$$

Suppose the conclusion p to be false: then $\sim p$ and $\sim q$ together contradict $p \vee q$. Therefore the supposition is wrong, and the conclusion is a true consequence of the premises.

Example 9. Show that the following argument is valid:

$$p \vee q \quad (i)$$
$$p \to r \quad (ii)$$
$$\sim b \quad (iii)$$
$$\overline{\quad r \quad}$$

Suppose $\sim r$ is true. Then:

$$\sim r$$
$$\frac{\sim r \to \sim p \quad \text{(contrapositive of } ii)}{\sim p \quad \text{(detachment)}}$$

We thus have $\sim p$ and $\sim q$, contradicting $p \vee q$: hence the supposition is false and the conclusion r is a true deduction.

Third method. We may show that the premises together with the assumption that the conclusion is false lead to a contradiction. This is a variation of the second method (compare the proof of Theorem I p. 36).

Example 10. Examine the validity of the argument:

$$p$$
$$\frac{(\sim p \vee \sim q) \to \sim p}{q}$$

Consider the two premises together with the negation of the conclusion:

$$p \quad (i)$$
$$(\sim p \vee \sim q) \to \sim p \quad (ii)$$
$$\sim q \quad (iii)$$

From (ii), (iii) we have $\sim p$, and this with (i) gives $p \wedge \sim p$ which is a contradiction: hence the negation of the conclusion is false, and the conclusion is a true deduction.

Exercise 9d

Give an indirect proof of each of the following propositions:
1. If the product of two integers is even, at least one of the integers is even.

2. $\sim p$
   $r \to p$
   $\frac{\sim q \vee r}{\sim q}$

3. q
   $\sim q \vee \sim r$
   $\frac{\sim p \to r}{p}$

4. $p \to \sim q$
   $r \leftrightarrow p$
   $\frac{q}{\sim r}$

5. $p \to q$
   $(q \wedge s) \to t$
   $\frac{s \wedge p}{t}$

114 SETS AND SYMBOLIC LOGIC

9.4 Verbal arguments

In testing mathematically the validity of a verbal argument, the first step is to express the premises and conclusion symbolically. When doing this, the following points should be noted:

1. It is advisable to let p, q, r . . . stand for positive simple statements (cf. Example 5, p. 87).
2. Each simple constituent statement should be denoted by a symbol. Example 3, p. 86, illustrates the danger of regarding a conjunction of two statements as a simple statement because the connective 'and' does not happen to appear.
3. The various ways in which a condition may be expressed should be recognized. Note the form 'p unless q' which means 'If not q, then p', *i.e.*, $\sim q \rightarrow p$.

Exercise 9e

Write symbolically:
1. p if q
2. p only if q
3. If q then p
4. p implies q
5. From p it follows that q
6. p is a sufficient condition for q
7. p is a necessary condition for q
8. Not p iff q
9. p unless not q
10. Not q unless not p

When an argument has been written symbolically, its validity may be tested by the methods of Sections 9.1, 9.2, 9.3.

Example 11. Examine the validity of the following argument:

Bill will study medicine or law	(*i*)
He will not study medicine unless his mark in biology is high	(*ii*)
Bill will not study law	(*iii*)

Bill gained a high mark in biology

Let p be 'Bill will study medicine'
  q be 'Bill will study law'
  r be 'Bill gained a high mark in biology'

Then the argument is:

$$p \vee q \quad (i)$$
$$\sim r \to \sim p \quad (ii)$$
$$\underline{\sim q \qquad (iii)}$$
$$r$$

If (ii) is replaced by its contrapositive, the argument becomes that of Example 9, p. 113, which has been shown to be valid.

Example 12. Examine the validity of the following argument for the conclusion 'The reward is not attractive.'

If the reward is attractive, then it is worth competing for. Either sufficient time is allowed, or the reward is not worth competing for. The time allowed is insufficient.

Let p be 'The reward is attractive'
    q be 'The reward is worth competing for'
    r be 'Sufficient time is allowed'.

Then the argument is:

$$p \to q \quad (i)$$
$$r \vee \sim q \quad (ii)$$
$$\underline{\sim r \quad (iii)}$$
$$\sim p$$

From (ii), (iii) we have $\sim q$. This with the contrapositive of (i) gives $\sim p$. Hence the argument is valid for the given conclusion.

Exercise 9f

Test the validity of the following arguments for the given conclusions (they are not all valid):

1. I enjoy my holiday if I feel well. I do feel well. Therefore I enjoy my holiday.
2. If the match is exciting the crowd is large. If the crowd is large the gate receipts are high. Therefore a match which is exciting is financially profitable.
3. Profits are high if and only if modern methods are employed. Structural improvements are made if profits are high. Structural improvements are not being made. Therefore modern methods are not being employed.

4. The candidate wins the election only if he gains people's confidence. He gains people's confidence unless he is self-opinionated. Therefore if the candidate wins the election he is not self-opinionated.

5. The picture is underpriced or it is a forgery. If the picture is underpriced then there is a rush to buy it. There is no rush to buy it. Therefore the picture is a forgery.

Quantified statements

The words 'all', 'some', 'none' sometimes occur in verbal arguments. Where they do, a verification of the argument by Venn diagrams may be appropriate.

Example 13. Consider the argument:

All sick people are weak	(*i*)
Nobody is foolish who can make money	(*ii*)
All weak people are foolish	(*iii*)

Sick people cannot make money

Premise (*i*) says that the set of sick people (S, say) is a subset of the set of weak people (W).

Premise (*iii*) says that the set of weak people is a subset of the set of foolish people (F).

Premise (*ii*) says that no member of the set of money-makers (M) belongs to set F; *i.e.*, M and F are disjoint.

The Venn diagram is therefore shown in Fig. 9.1. It shows the argument to be valid since the sets S and M are necessarily disjoint.

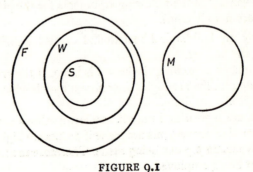

FIGURE 9.1

Example 14. Consider the argument:

All artists are men of insight (*i*)
Only men without insight become politicians (*ii*)
Some musicians are politicians (*iii*)

Some musicians are not men of insight

Premise (*i*) says that the set of artists (*A*) is a subset of the set of men of insight (*I*).

Premise (*ii*) says that *I* and the set of politicians (*P*) are disjoint.

Premise (*iii*) says that if *M* is the set of musicians, *M* and *P* are not disjoint, *i.e.*, $M \cap P \neq \emptyset$. We have no information as to whether *I* and *M* are disjoint, but this does not matter. Fig. 9.2 shows that

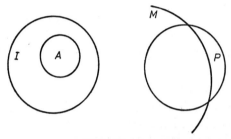

FIGURE 9.2

some elements of *M* are not elements of *I* which proves the argument valid for the given conclusion.

Had the conclusion been 'Some musicians are artists' the argument would have been fallacious, since we cannot deduce from the premises that the sets *M* and *A* have common elements.

9.5 The laws of symbolic logic

The laws of symbolic logic (or 'the algebra of propositions') may be summarized as were the laws of the algebra of sets (Section 6.2, p. 71). They will all have been proved by the student who has worked through the exercises of Chapters 7–9.

A. Primary laws

1. $p \vee p = p$ 2. $p \wedge p = p$
3. $p \vee f = p$ 4. $p \wedge t = p$
5. $p \vee t = t$ 6. $p \wedge f = f$

B. *Associative laws*

1. $(p \vee q) \vee r = p \vee (q \vee r)$
2. $(p \wedge q) \wedge r = p \wedge (q \wedge r)$

C. *Commutative laws*

1. $p \vee q = q \vee p$
2. $p \wedge q = q \wedge p$

D. *Distributive laws*

1. $p \vee (q \wedge r) = (p \vee q) \wedge (p \vee r)$
2. $p \wedge (q \vee r) = (p \wedge q) \vee (p \wedge r)$

E. *Laws of negation*

1. $p \vee \sim p = t$ 2. $p \wedge \sim p = f$
3. $\sim f = t$ 4. $\sim t = f$
5. $\sim (p \vee q) = \sim p \wedge \sim q$ 6. $\sim (p \wedge q) = \sim p \vee \sim q$
7. $\sim (\sim p) = p$

As in the algebra of sets, the distributive laws may be written with distribution from left or right in virtue of the commutative laws. Disjunction is left- and right-distributive over conjunction; and conjunction is left- and right-distributive over disjunction.

Laws E5 and E6 are the application of De Morgan's laws to the algebra of propositions.

Exercises on Chapter 9

Examine mathematically the validity of the following arguments for the conclusions stated:

1. If my brother eats cake he does not take sugar in tea. He does not eat an egg unless he has toast as well. He eats toast only if he has sugar in tea. My brother had a piece of cake today. Therefore he had no egg today.

2. If old people are feeble then long life is undesirable. That the world is a happy place implies that long life is desirable. It is true that the world is a happy place and that old people are feeble. Therefore the world is both a happy and an unhappy place.

3. It is false that a fine June means a wet July. If July is dry, then

there is a water shortage. If there is an influx of tourists there is a water shortage. It is a fine June. Therefore the tourist trade is booming.

4. When taxes do not rise people are happy. If people are happy there is no emigration. Many people are emigrating. Therefore taxes are rising.

5. If roses do not bloom freely the soil needs enriching. When roses are badly pruned they do not bloom freely. If fertilizer is used the soil does not need enriching. Therefore if roses are not well pruned fertilizer has not been used.

6. Chinese is a difficult language and teaching is a rewarding profession. That Chinese is a difficult language implies that if teaching is rewarding then Chinese sounds musical. Therefore Chinese sounds musical.

7. Bill is thirsty if he plays tennis. He drinks lime juice when he is thirsty. If he drinks lime juice he feels ill. Therefore Bill feels ill if he plays tennis.

8. Some birds can talk. No talkers are uninteresting. Therefore all birds are interesting.

9. All athletes are hard-working, good-looking, or well-spoken. All hard-working people are well-spoken or clear-thinking. All well-spoken people are good-looking and clear-thinking. Some good-looking people are clear-thinking. Therefore some athletes are clear-thinking.

10. All newspapers are intended for mass-production. Nothing mass-produced can be attractive. Therefore no newspaper is attractive.

11. Some boxes are cases. Some containers are boxes. No cases are containers. Therefore some boxes are neither cases nor containers.

12. Some composers' work gives pleasure. Nothing that gives pleasure is worthless. Some composers' work is inspired. Therefore some inspired compositions are worthless or give pleasure.

10 | INTRODUCTION TO BOOLEAN ALGEBRA

10.1 Introduction

In this book we have been studying two 'algebras', that is, two mathematical systems with operations governed by a set of laws. We have seen that the algebra of sets and the algebra of propositions are closely related: and in this chapter we shall show that they are both applications of an abstract system which is called Boolean algebra, after the English logician George Boole (1815–64). Boolean algebra is more fully developed in 'Mathematical Structures' in this series.

10.2 Boolean algebra

In a Boolean algebra we have:

(I) a set of elements
(II) two binary operations
(III) four laws

(I) We will denote the set by B and its elements by $a, b, c, \ldots$
(II) A binary operation was defined in Section 3.1, p. 27. We denote the two binary operations of Boolean algebra by the symbols $(+)$ and $(.)$; and, instead of writing $a.b$, we write ab.

It is most important to remember that these familiar signs do not here denote sum and product as they do when applied to numbers. In Boolean algebra, $a+b$ may be read as 'a plus b', but a and b are not numbers, and $a+b$ does not denote a simple sum: nor does ab denote a simple product. This giving of new meanings to familiar symbols makes for economy, and ought not to cause confusion if you give due thought to what you are doing.

(III) The four laws of Boolean algebra are:

1. The operations denoted by (+) and (.) are commutative. Thus:

   (i) $a + b = b + a$ (ii) $ab = ba$

2. Each operation is distributive over the other. Thus:

   (i) $a(b + c) = ab + ac$
   (ii) $a + bc = (a + b)(a + c)$

3. Each operation has an identity element associated with it.
   (i) For (+) we denote the identity element by 0. Thus:

   $$a + 0 = a$$

   (ii) For (.) we denote the identity element by 1. Thus:

   $$a.1 = a$$

   From the first law it follows that $0 + a = a$ and $1.a = a$.

4. For every element a of the set B there exists an *inverse element* a' of B defined by:

   (i) $a + a' = 1$ (ii) $aa' = 0$

   The dash (') is sometimes referred to as *prime*.

Two points about these laws should be noted:

(a) In law 2, we have stated the left-distributive law. It is not necessary also to state the right-distributive law: why?

Observe that the law of distribution of (.) over (+) leads to the familiar equation $a(b + c) = ab + ac$, whereas the law of distribution of (+) over (.) gives $a + bc = (a + b)(a + c)$, which probably looks wrong. If a, b, c were numbers, and (+), (.), had their original meanings of sum and product this equation would not hold. In Boolean algebra it does.

(b) In law 3, although the symbols 0 and 1 are used for identity elements, and may be read as 'zero' and 'one', they must not be thought of as integers. They also are familiar symbols to which new meanings are given.

10.3 The algebra of sets as a Boolean algebra

To show that the algebra of sets is a Boolean algebra, it is only necessary to show that it satisfies conditions (I), (II), (III) which define a Boolean algebra.

(I) Sets P, Q, R, ... correspond to the elements a, b, c, ... of Boolean algebra.

(II) In set theory there are two binary operations denoted by $\cup$ and $\cap$: these correspond to the Boolean operations denoted by $(+)$ and $(.)$, respectively.

(III) There is an exact correspondence between the four laws of Boolean algebra and certain of the laws of sets. The following table shows this correspondence: it should be carefully studied. Note that 0 corresponds to $\emptyset$, and 1 to $\mathscr{E}$. The references are to Sections 10.2 and 6.2 (p. 71).

Boolean algebra		Algebra of sets	
$a+b = b+a$	[1(i)]	$P \cup Q = Q \cup P$	[C1]
$ab = ba$	[1(ii)]	$P \cap Q = Q \cap P$	[C2]
$a(b+c) = ab+ac$	[2(i)]	$P \cap (Q \cup R) = (P \cap Q) \cup (P \cap R)$	[D2]
$a+bc = (a+b)(a+c)$	[2(ii)]	$P \cup (Q \cap R) = (P \cup Q) \cap (P \cup R)$	[D1]
$a+0 = a$	[3(i)]	$P \cup \emptyset = P$	[A3]
$a.1 = a$	[3(ii)]	$P \cap \mathscr{E} = P$	[A4]
$a+a' = 1$	[4(i)]	$P \cup P' = \mathscr{E}$	[E1]
$a.a' = 0$	[4(ii)]	$P \cap P' = \emptyset$	[E2]

There are other laws of the algebra of sets not quoted in this table: it is possible to show that each of them is satisfied by a Boolean algebra. This is developed in 'Mathematical Structures' in this series: one will be proved here as an illustration.

Example 1. State and prove a Boolean identity corresponding to the set law $P \cup P = P$ [A1].

The corresponding Boolean identity is $a+a = a$.

$$
\begin{aligned}
\textit{Proof.} \quad a &= a+0 & [3(i)] \\
&= a+aa' & [4(ii)] \\
&= (a+a)(a+a') & [2(ii)] \\
&= (a+a).1 & [4(i)] \\
&= a+a & [3(ii)] \\
\text{whence} \quad a+a &= a & [1(i)]
\end{aligned}
$$

10.4 Dual theorems

Note that the principle of duality in the algebra of sets (Section 6.3, p. 72) applies also in Boolean algebra. If in any Boolean theorem the operations denoted by (+) and (.) are interchanged and also the identity elements 0 and 1, another theorem, the dual of the first, will emerge.

Thus the dual of the theorem $a + a = a$, just proved, is $aa = a$. State the law of the algebra of sets which corresponds to this theorem.

Example 2. Prove the theorem $a + a'b = a + b$, and write its dual.

$$\text{Proof.} \qquad a + a'b = (a + a')(a + b) \qquad [2(ii)]$$
$$= 1.(a + b) \qquad [4(i)]$$
$$= a + b \qquad [3(ii)]$$

The dual is $\qquad a(a' + b) = ab$

Exercise 10a

1. Write theorems in the algebra of sets corresponding to the following Boolean identities:

   (a) $a + ab = a$ (c) $(a + b)(a' + c) = ac + a'b$

   (b) $(ab)' = a' + b'$ (d) $ab + a' + b' = 1$

   Verify your answers by means of Venn diagrams.

2. Write Boolean theorems corresponding to the following set theorems:

   (a) $(P \cup Q)' = P' \cap Q'$ (c) $P \cup (P \cap Q) = P$

   (b) $(P \cup \mathscr{E}) \cap (P \cap \emptyset) = \emptyset$ (d) $(P - Q) \cap Q = \emptyset$

3. Write the duals of:

   (a) $ab + a' + b' = 1$

   (b) $(a.0)(a + 1) = 0$

   (c) $ac + a'b + bc = ac + a'b$

   (d) $ab + bc + ca = (a + b)(b + c)(c + a)$

4. Prove, using the laws of Boolean algebra:

   (a) $a(a' + b) = ab$ (c) $a'b'(a + b) = 0$

   (b) $a + (a + 1) = 1$ (d) $ab + a' + b' = 1$

10.5 The algebra of propositions as a Boolean algebra

A table similar to that of Section 10.3 might be drawn up to show that each of the laws of Boolean algebra may be expressed in terms of statements p, q, r, ... with $\vee$ and $\wedge$ replacing $(+)$ and $(.)$, and f, t replacing 0, 1.

Exercise 10b

Draw up such a table, giving full references to the laws of the algebra of propositions (Section 9.5, p. 117).

Exercises on Chapters 1–10

1. By finite induction prove that the nth number in the sequence 1, 3, 7, 15, 31, ... is $2^n - 1$. (Each number of the sequence is one greater than twice the preceding number.)
2. Is p a necessary condition for $p \wedge q$? Is p a sufficient condition for $p \wedge q$?
3. If P, Q are the truth sets of statements p, q write down the statement whose truth set is $P - Q$. Draw up its truth table, and identify the rows of the table with regions in the Venn diagram of $P - Q$.
4. Let P, Q, R be three sets. Prove that:
$$n(P \cap Q) = n(P \cap Q \cap R) + n(P \cap Q \cap R')$$
5. Prove that $(3, 5).(6, 2) = (5, 3).(2, 6)$ where the ordered pairs represent integers. Is the relation true where the ordered pairs represent rationals?
6. Let $\sim p \to q$ be denoted by $p \uparrow q$.
   (i) Express in terms of $\uparrow$:
       (a) $p \vee q$ (c) p unless q
       (b) p only if q (d) $q \to \sim p$
   (ii) Prove $(p \uparrow \sim q) \wedge (\sim p \uparrow q) = p \leftrightarrow q$.
7. Defining $p \uparrow q$ as in the previous question, and $p \downarrow q$ as $p \to \sim q$;
   (a) prove that $p \uparrow q$ is the converse of the contrapositive of $p \downarrow q$.
   (b) is $(p \uparrow q) \to (\sim q \downarrow \sim p)$ a tautology?
   (c) draw up the truth table of $(p \downarrow q) \wedge \sim (q \uparrow p)$.

8. Prove that $a + 1 = 1$, where a is an element in a Boolean algebra.

9. Using any valid method, replace by a simpler equivalent the proposition $[p \lor (q \leftrightarrow \sim p)] \to \sim q$.

10. By means of a truth table, prove that $P \cup Q \cup (P' \cap Q') = \mathscr{E}$ where P and Q are any two sets.

11. Illustrate the equivalence $p \to q = \sim p \lor q$ by re-phrasing the following conditional statements in disjunctive terms:
    (a) If the clock is slow we shall miss the train.
    (b) If $AB = AC$, then $A\hat{B}C = A\hat{C}B$.
    (c) If you exceed 20 km/h I shall report you to the police.
    (d) For the venture to succeed it is necessary that more capital be invested in it.
    (e) Industry is a sufficient condition for success.
    (f) To pass the examination it is enough to answer five questions correctly.

12. Examine the validity of the following argument for the stated conclusion:

> No one under 21 may join the club. All capable people are busy. Some club members are busy. Therefore not all capable people are club members.

13. Show that $(2 + x)^{13} = 2^{13}(1 + \frac{1}{2}x)^{13}$. In the binomial expansion of $(2 + x)^{13}$ what are the coefficients of:
    (a) x^4 (b) x^{10}

14. A committee of 5 is to be chosen from 9 men with the condition that two of the men, A and B, must not be on a committee together. In how many ways may the committee be selected?

15. p and q are two inconsistent statements (i.e., $p \land q = f$). If P, Q are their truth sets and $n(P) = 15$, $n(Q) = 24$, how many elements are there in the truth set of $p \lor q$?

16. Let $p \updownarrow q$ denote $\sim(p \leftrightarrow q)$.
    (a) Show that the truth set of $p \updownarrow q$ is $P \triangle Q$ where P, Q are the truth sets of p, q.
    (b) Draw up a truth table for $(\sim p \land q) \to (p \updownarrow q)$. What does the result show?

17. Simplify:

    (a) $\dfrac{n! - (n-1)!}{(n-1)}$ (b) $\dbinom{2n}{n} \cdot \dbinom{3n}{2n} \cdot \dbinom{4n}{3n}$

18. Is the operation denoted by $\updownarrow$ (question 16) commutative? Is it associative?

19. Figure 10.1 shows a maze used in experiments with mice. The mouse enters at A, and obtains food at G. Draw a tree diagram showing the eight different routes it may take in going from A to G. In how many routes does it make exactly two mistakes? (Assume that in correcting each mistake it moves nearer to G.)

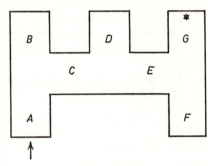

FIGURE 10.1

20. A and B are two sets, and the product set $A \times B$ has twelve elements. What are the possibilities for the number of elements in A, B?

21. Test the validity of the argument: if $p \lor q$, then $p \to \sim q$.

22. Write out the solution set of the equation:
$$x(x-1)(2x+1)(x^2-2) = 0$$
    (a) where $x \in N$ (c) where $x \in Q$
    (b) where $x \in Z$ (d) where $x \in R$

23. Prove that $2^{2n+1} + 1$ is divisible by 3 for all $n \in N$. Is there any other integral value of n for which this is true?

24. By means of a counter example prove that the following statement is false: 'For all $n \in N$, the number $n^2 - n + 41$ is prime.'

25. Let the set operation ϕ be defined by $P \phi Q = P' \cup Q'$.
    (a) Is ϕ a commutative operation?
    (b) Is ϕ an associative operation?

26. Is $p \lor q$ a necessary condition for $p \land q$? Is $p \lor q$ a sufficient condition for $p \land q$?

27. Four people A, B, C, D reported what they allegedly saw at the scene of an accident.
A said:
'Some people in the red car were hatless and all wore beards.'
B said:
'There were some in both cars wearing hats and glasses.'

C said:

'None of the occupants of the white car had a beard.'

D said:

'No one with a hat wore glasses unless he had a beard.'

Would you accept the testimony of these four in its entirety? Give valid reasons for your decision.

28. A *diophantine* equation in two unknowns, x and y, is an algebraic equation whose coefficients and solutions are integers. Prove that the diophantine equation $3x + 5y = 22$ has the solution set $\{(x, y) | x = 4 - 5n, y = 3n + 2, n \in Z\}$.

29. Find the solution set of the diophantine equation $6x - y = 11$.

30. (a) If $(p \wedge q)$ and $p \leftrightarrow q$ are both false, what can be said about $(p \vee q)$?

(b) If $p \to q$ is true and $p \leftrightarrow q$ is false, what can be said about:

(i) $(p \wedge q)$ (ii) $(p \wedge q) \vee (p \vee q)$

31. Let $p \, \psi \, q$ denote $(p \vee q) \wedge \sim(p \to q)$. Prove:

(a) ψ is not a commutative operation

(b) $(p \to q) \, \psi \, q = \sim(p \vee q)$

32. How many integers are factors of $3.5^2.8^3.11^3$? (Include 1 and the number itself.)

33. Name nine elements (x, y) of the set:

$$S = \{(x, y) | x, y \text{ and } x + y \text{ are positive primes}\}$$

34. Prove by finite induction:

$$1^3 + 3^3 + 5^3 + \ldots + (2n + 1)^3 = (n + 1)^2(2n^2 + 4n + 1)$$

35. (a) Find the maximum number of points of intersection of 8 straight lines.

(b) An *n*gon is an *n*-sided polygon: a diagonal is a line joining two non-consecutive vertices. How many diagonals has an *n*gon?

36. A group of 160 people, Americans, Germans and Frenchmen, were asked whether they approved of making Esperanto compulsory in schools. Their replies are tabulated:

	No	Yes	Undecided
Americans	23	4	8
Germans	30	23	9
Frenchmen	43	17	3

If N is the set of non-Germans, E is the set of Europeans, Y is the set of those who said 'Yes' and U is the set of those who were undecided, find:

(a) $n(N \cap Y)$

(b) $n[(E - N) \cap U]$

(c) $n[E \cap (Y \cup U)']$

Express symbolically:

(d) the set of Frenchmen interviewed

(e) the set of Americans interviewed

(f) the set of Germans who said 'No'

37. The binary operation $\circ$ is defined by $P \circ Q = R$, where P and Q are any two points in a plane and R is the mid-point of PQ. Investigate whether $\circ$ is commutative and/or associative.

(AEB)

Topics for investigation

Sheffer's stroke

About 50 years ago, the logician H. M. Sheffer pointed out that all five connectives $\sim$, $\wedge$, $\vee$, $\rightarrow$, $\leftrightarrow$ can be expressed in terms of a single symbol, which has come to be known as Sheffer's stroke.

If p/q stands for $\sim(p \wedge q)$, then p/p will be $\sim(p \wedge p)$, i.e., $\sim p$. Thus p/p is another way of writing the negation of p. Since p/q is the negation of $(p \wedge q)$, it follows that $(p/q)/(p/q) = p \wedge q$.

Investigate how $\vee$, $\rightarrow$ and $\leftrightarrow$ can be expressed in terms of Sheffer's stroke.

Logical brain-teasers

Lewis Carroll was the pen-name of Charles Lutwidge Dodgson (1832–98), the Cambridge mathematician who wrote 'Alice in Wonderland' and 'Alice Through the Looking-Glass'. He also wrote, among other books, 'Symbolic Logic' and 'The Game of Logic', now obtainable as one volume (published by Dover Publications Inc., and distributed in the UK by Constable & Co. Ltd).

If you can consult this in the library, you will find many amusing examples of arguments to which to apply the mathematical methods of testing validity which you have been studying. Here is one example from Lewis Carroll:

All ducks in this village which are branded B belong to Mrs Bond.

Ducks in this village never wear lace collars, unless they are branded *B*.

Mrs Bond has no gray ducks in this village.

Is the conclusion 'No gray ducks in this village wear lace collars' valid?

Doubtful evidence

In a certain society all full members are invariably truthful, while no associate member is ever truthful. A visitor talking to three members *A*, *B*, *C* asked *A* if he was a full member. When *A* had replied, *B* said that *A*'s answer was 'Yes'. *C* then said that actually *A* was an associate member. How many of the three were full members?

How is this teaser constructed to ensure that it has a precise answer? Try to devise some similar puzzles.

The Fibonacci sequence

Obtain a sheet of foolscap paper and, if necessary, trim it so that the measurements are in the ratio of $13:8$. On it draw a line dividing it into a square and a rectangle. Then draw a line dividing this new rectangle in the same way. Continue in this way with each new rectangle until you finally obtain one which you can divide into two squares.

Denote one of these squares by *A*, the rectangle from which it was obtained by *B*, and the other rectangles by *C*, *D*, *E*, *F* in increasing order of size, *F* being the complete sheet.

Complete this table:

Figure	Breadth, x	Length, y
A		
B		
C		
D		
E		
F	8	13

The values of x in this table are the first six elements of the Fibonacci sequence. Note that the first two terms are both unity. Write down the next six elements, and state a rule by which any element of the sequence after the second may be derived from preceding elements.

A rectangle has often been regarded as having the most pleasing proportions if the square of its breadth (x) is equal to the product of its length (y) and the difference between its length and breadth, *i.e.*, if $x^2 - y(y - x) = 0$. (See what information you can obtain about this: it may be referred to as 'golden section' or 'dynamic symmetry'.)

Examine whether rectangles whose breadth and length are consecutive terms of the Fibonacci sequence approximate to this ideal by completing the table:

x	y	x^2	$y(y-x)$	$x^2 - y(y-x)$
1	1	.	.	.
1	2	.	.	.
2	3	4	3	1
.	.	.	.	.
.	.	.	.	.
8	13	64	65	−1
.	.	.	.	.
.	.	.	.	.
.	.	.	.	.
34	.	.	.	.

State in terms of x and y the relation which you observe from this table.

If x, y, z are any three consecutive elements of the Fibonacci sequence, what is the value of z in terms of x and y?

Arising from this, prove algebraically that if x, y are related by the equation $x^2 - y(y - x) = 1$, then y, z are related by the equation $y^2 - z(z - y) = -1$.

If a rectangle is of length of 13 cm, and the ideal conditions are fulfilled, what is its breadth correct to two decimal places? (Take $\sqrt{845}$ as 29·07.) What does this result show about the properties of a sheet of foolscap paper (13 in × 8 in)?

Find out what you can about other interesting applications of the Fibonacci sequence.

The type area of this page (*i.e.*, the distance from the top of the running headline to the bottom of the last line times the width of a full line) measures 172 mm × 106 mm. Check that this is a close approximation to a 'golden' rectangle.

ANSWERS

Exercise 1a
1. $\notin$ 2. $\in$ 3. $\notin$ 4. $\notin$ 5. $\in$

Exercise 1b
1. $A = \{$Feb, Apr, June, Sep, Nov$\}$
   $B = \{1, 2, 4, 5, 7, 8\}$
   $C = \{B, C, D, E, H, I, K, O, X\}$
   $D = \{POQ, QOR, ROS, SOP,$
   $PQR, QRS, RSP, SPQ\}$
   $E = \{2, 4, 6, 8, 10\}$
2. $F = \{n \,|\, n = x^2,\; x$ is an integer,
   $-7 < x < 7\}$
   $G = \{n \,|\, n = 3x - 1,\; x$ is a positive
   integer$\}$
   $H = \{x \,|\, x$ is a cardinal point of the
   compass$\}$
   $K = \{x \,|\, x = \sqrt{y},\; y$ is a positive
   integer, $y < 10\}$
   $L = \{x \,|\, x = \dfrac{1}{y},\; y$ is a positive
   integer, $y < 21\}$
3. X is a geometrical progression:
   first term, 1; common ratio, 2
   Y is the set of multiples of 5
   Z is the set of the three school
   science subjects

Exercise 1c
1. $A = B$ 3. $A = B$
2. $A \neq B$ 4. $A \neq B$

Exercise 1d
1. (a) $T \subset P$ (b) $Q \subset R$ (c) $R \supset Q$
   (d) $Q \subset S$ (e) $S \supset R$
2. $B \subset A,\; D \subset A$. No
3. $A \subset B, C \supset A, B \supset A$. Neither is
   a subset of the other; $B \not\subset C$,
   $C \not\subset B$
4. (a) $A \subset C$ (b) $C \subset L$
   (c) $B \not\subset L$ (d) $L \supset A$

Exercises on Chapter 1
1. (a) $\{4\}$ (b) $\{-11, 0, 1 \cdot 5\}$ (c) $\emptyset$
   (d) $\{-\tfrac{1}{3}, \tfrac{1}{3}\}$ (e) $\mathscr{E}$
   (f) $\{-3, -1\}$ (g) $\{1, 4\}$
2. S_3 has $2^3 = 8$ elements; S_4 has
   2^4, and S_n has 2^n
3. Only (b) is false
5. $\emptyset$
7. A and B are possibly, but not
   necessarily, comparable
8. $B \subset A,\; B \subset C,\; B \subset D,\; D \subset C$.
   A and C are non-comparable; so
   are A and D
9. $A = \{\tfrac{1}{6}, \tfrac{2}{6}, \tfrac{3}{6}, \tfrac{4}{6}, \tfrac{5}{6}\}$
   $B = \{1, 2, 3, 4, 6, 8, 9, 12, 18,$
   $24, 36, 72\}$
   $C = \{7, 14, 21, 28, 35, 42, 49\}$
   $D = \{$sight, hearing, smell, taste,
   touch$\}$
   $E = \{5, 8, 2, 6, 9\}$
10. (a) is the set of all real numbers
    except zero
    (b) is $\mathscr{E}$ (c) is $\emptyset$ (d) is $\{1\}$
11. $\{3, 4, 7, 8\}$ $\{3, 7, 8\}$ $\{4, 7, 8\}$
    $\{3, 4, 8\}$ $\{7, 8\}$ $\{8\}$ $\{3, 8\}$ $\{4, 8\}$

Exercise 2a
1. (a) $\{-4, -2, 0, 2, 4\}$
   (b) $\{-3, -2, -1, 1, 2, 3\}$
   (c) P
2. $A \cup B = \{p, r, o, c, e, s, t\}$
3. (a) $\{a \,|\, a = \tfrac{1}{n},\; n$ is a positive in-
   teger, $n < 11\}$
   (b) $\{x \,|\, x = 2n,\; n$ is a positive in-
   teger, $n < 7\}$
   (c) $\{x \,|\, (3x - 1)(3x - 4)(x - 2) = 0\}$

Exercise 2b
2. (a) Both are A (or B)
   (b) $A \cap B = A$ (c) $\emptyset$
   (d) $A \cap \mathscr{E} = A$

4. Yes
5. Yes   Yes
6. (a) {4, 6, 8}
   (b) {1, 2, 3, 4, 5, 6, 7, 8, 10}
   (c) $\{x \mid x$ is a positive integer, $x < 10\}$
   (d) {1, 2, 3, . . ., 8}
   (e) $\{x \mid x$ is a positive integer, $x < 9\}$

Exercise 2c

1. (a) {2, 4, 5, 6, 8}
   (b) {2, 3, 4, 6, 7, 8}
   (c) {1, 3, 5, 6, 7, 9}
   (d) {2, 4, 8}
   (e) $\mathscr{E}$
   (f) {5, 6}
   (g) {6}

Exercise 2d

1. (a) {3, 7}   (b) {5, 6}   (c) {5}
   (d) {1, 9}   (e) A (f) $\emptyset - A = \emptyset$

Exercise 2e

1. (a) {3, 5, 7}   (b) {5, 6}
   (c) {1, 2, 4, 8, 9}   (d) {3, 5, 7}
   (e) {6}   (f) {3, 5, 7}
2. (a) {p, w}   (b) {p}
   (c) {p, v}   (d) {p, q, r, w}

Exercises on Chapter 2

1. (a) {1, 2, 3, 4, 11}
   (b) {12, 13, 14, 15}
   (c) $\emptyset$
   (d) {1, 2, 3, 4, 5, 12, 13, 14, 15}
2. *fruition* belongs to B, C, A'; *amplitude* belongs to A', B', C'; $A \cap C = \emptyset$
3. $D_{24} \cap D_{54} = \{1, 2, 3, 6\} =$ the set of factors of the greatest common divisor of 24 and 54.
6. Sets (a) and (b) are equal, suggesting that intersection distributes over symmetric difference. Symmetric difference does not distribute over intersection

7. Union does not distribute over symmetric difference, nor does symmetric difference distribute over union
8. (i) (a) adults
       (b) unmusical females
       (c) musical males
       (d) musical girls
   (ii) (a) $M' \cap C \cap F'$
        (b) $M \cap C' \cap F'$
        (c) $(M \cap C) \cup (M' \cap C')$
12. (i) (a) {3, 4, 5}
        (b) {8, 9}
        (c) {3, 4, 5}
    (ii) $\{x \mid 2 < x < 6, 7 < x < 10\}$

Exercises on Chapters 1 and 2

4. {4, 6, 9}
6. (a) $(B - A) - C$
   (b) $[(A - B) \cup C]'$
   (c) $(B \cup C) - A$
   (d) $(B - A') \cup (C - A)$
   (e) $A - (B \cup C)$
   (f) $B' - C$
8. (a) {B}
   (b) The set of points on the bisector of $A\hat{B}C$; the set of points on the bisector of $B\hat{C}A$
   (c) $S_P \cap S_Q$ is the set whose single element is the centre of the inscribed circle of ABC
   (d) The set of points of contact of the inscribed circle

Exercise 3a

2. No
3. {c, b, f, e, a, d}
4. No
5. Yes

Exercise 3c

2. $k = 4$
4. (a) (1, 7)   (b) (12, 1)   (c) (1, 8)

Exercise 3d

1. (3, 1)   2. (2, 1)   3. (1, 3)   4. (1, 3)

Exercise 3e
2. False 3. Yes

5.

x	1	2	3	4	6	8	12	24
y	24	12	8	6	4	3	2	1

Exercise 3f
1. (a) 6   (b) -5 (c) 6
   (d) -1 (e) 0   (f) -9
2. (a) -7 (b) -12 (c) 0

Exercises on Chapter 3
1. Yes 2. No. No
6. (a) $\{-0.6, 0.6\}$ (b) $\{-7\}$
   (c) $\emptyset$ (d) $\emptyset$ (e) $\{1\}$ (f) $\{1\}$

Exercises on Chapters 1-3
1. $M_2 \,\Delta M_3 = \{x \,|\, x = 6n - 4, 6n - 3,$
   $6n - 2, \ n \in N\}$. Lower, $6n - 3$;
   higher, $6n + 2$
3. (a) θ is not commutative: $\emptyset$ is
     commutative
   (b) θ is not associative; $\emptyset$ is
     associative
   (d) θ is not distributive over $\emptyset$

7.

x	4	8	12
y	12	7	2

12. X is closed under multiplication

Exercise 4a
2. Both are $\{(a, b), (b, b)\}$
3. $\times$ is distributive over $\cup$
4. (a) $\{(1, 1), (1, 3), (2, 1), (2, 3)\}$
   (b) $\{(1, 2), (1, 3), (2,2), (2, 3)\}$
   (c) $\{(1, 1), (2, 1), (1, 2), (2, 2)\}$
   (d) $\{(3, 1), (3, 3)\}$
   (e) $\emptyset$

Exercise 4b
1. $P = \{x \,|\, x \in R, \ 1 \leqslant x \leqslant 2\}$
   $Q = \{y \,|\, y \in R, \ -3 \leqslant y \leqslant -1\}$
2. $\{(p, p, p,) \ (p, p, f), \ (p, f, p),$
   $(f, p, p), (p, f, f), (f, p, f), (f, f, p),$
   $(f, f, f)\}$

Exercises on Chapter 4
7. $S = \{(-13, 0), (-12, -5), (-12,$
   $5), (-5, -12), (-5, 12), (0, -13),$
   $(0, 13), (5, -12), (5, 12), (12, -5),$
   $(12, 5), (13, 0)\}$
9. Fish is chosen in 6. Soup is
   followed by chicken in 3
10. Closure of s_2

Exercise 5a
1. A, D, E, F
2. Zero must be included as an
   element of A

Exercise 5b
1. Each is $\{\{a\}, \{b\}, \{c, e\}, \{d\}\}$
2. $\{\{F\}, \{H\}, \{K\}, \{G\}, \{J\}, \{L\}\}$

Exercise 5c
1. 4   2. 9   3. 3   4. 2
6. $n(P \cup Q) \not< 7$ and $\not> 12$
   $n(P \cap Q) \not> 5$

Exercise 5d
1. (a) 720   (b) 40320   (c) 1
2. (a) 7   (b) 56   (c) $n(n-1)$
3. (a) 5 !   (b) $(a-2)$!
5. 36.4 !

Exercise 5e
1. (a) $\dfrac{11!}{7!}$ (b) $\dfrac{x!}{(x-y-1)!}$
   (c) $\dfrac{8!}{5!\,3!}$
2. 120   3. 60 480   4. 6720

Exercise 5f
1. (a) 3   (b) 210   (c) $\frac{1}{2}n(n-1)$

Exercise 5g
1. 1680   2. 3
3. (a) 1260   (b) 3360
4. One partition; n ! ordered partitions

Exercise 5h
1. $1 + 4x + 6x^2 + 4x^3 + x^4$;
   $1 + 5x + 10x^2 + 10x^3 + 5x^4 + x^5$

2. (a) 120   (b) 252. The coefficients of x^3 and x^7 are the same. No. Six
3. 1716. Yes: the next term
4. The coefficient of x^{n-3} is:
$$\frac{n!}{(n-3)!\,3!}$$

Exercises on Chapter 5
2. $P' \cap Q'$
3. (a) 50   (b) 30   (c) 10   (d) 38
   (e) $M' \cap A$ or $(Y - M) \cap A$
   (f) $F' \cap (M - Y')$
   (g) $F \cap (Y' \cup M')$
   (h) $Y' \cap (F \cup A)'$
5. 87
7. 4; 6; 2
8. 48
9. (a) $\dfrac{18!}{3(6!).3!}$ (b) $\dfrac{18!}{6(3!).6!}$
10. (a) 36   (b) 15   (c) 30
11. 252
12. (a) 8   (b) 3   (c) 7

Exercise 6b
1. (a) $(A \cup B') \cup B = \mathscr{E}$
   (b) $(A \cap \emptyset) \cup (A \cup \mathscr{E}) = \mathscr{E}$
   (c) If $A \cup B = \mathscr{E}$, then
     $A' \cap B = A'$
   (d) If $(A \cup B') = A \cup B$ then
     $A = \mathscr{E}$
   (e) $A \cap B = (A \cup B) \cap$
     $(A \cup B') \cap (A' \cup B)$
   (f) $(A \cap \mathscr{E}) \cap (\emptyset \cup A) = \emptyset$
   (g) $(A \cup B) \supset (A \cap B)$

Exercises on Chapter 6
1. $P \cap (P' \cup Q) = P \cap Q$
2. $(X \cap \emptyset) \cup (X \cup \mathscr{E}) = \mathscr{E}$
3. $(A \cup B) \cap (B \cup C) \cap (C \cup A)$
   $= (A \cap B) \cup (B \cap C) \cup (C \cap A)$
4. $(P \cup R) \cap (P' \cup Q)$
   $= (P \cap Q) \cup (P' \cap R)$
5. $Q \subseteq P'$

14. (a) $(P' \cap Q') \cup (P \cap Q)$
   (b) $A \cap (B' \cup C)$
   (c) $P \cap Q'$

Exercises on Chapters 1–6
9. $(x, x), (x, z), (y, x), (y, y),$
   $(z, x), (z, y), (z, z)$
10. (a) $(P' \cap \mathscr{E}) \cup (Q' \cap P)$
     $\cup (P' \cap Q \cap \emptyset)$
   (b) $(P' \cup Q) \cap (P \cup Q')$
   (c) $P' \cap (Q' \cup R') \cap (Q \cup R)$
     $\cap (R' \cup \mathscr{E})$
   (d) $(P' \cup Q) \cap (P \cup Q')$
14. 4095
15. 5

Exercise 7a
(a), (e), (h), (j) are statements: only (a) and (j) are true.

Exercise 7b
(a) T (b) F (c) T (d) T
(e) T (f) T (g) F (h) F

Exercise 7c
1. (a) $\sim(p \wedge q)$ (b) $\sim(p \vee q)$
   (c) $p \vee q$ (d) $p \wedge (q \vee r)$
   (e) $(p \wedge q) \vee \sim r$
2. (b) The fuel supply is adequate and the heating system is in good order, but the house is not warm

Exercise 7d
1. (a) T (b) T (c) T (d) F (e) T
2. (b) If I work hard but do not take sufficient recreation I shall not succeed in my career
3. (a) $\sim(p \wedge q)$ (b) $\sim r \wedge \sim p \wedge q$
   (c) $r \to q$ (d) $\sim r \to (p \wedge q)$
   (e) $(r \wedge \sim q) \to \sim p$

Exercises on Chapter 7
2. (a) No   (b) Yes   (c) Yes
   (d) Yes   (e) No

3. (i)(b) An island is a piece of land entirely surrounded by water.
   (ii)(e) John is in the same Form as his grandmother
4. (a) F (b) F (c) T
   (d) F (e) F (f) T (g) T
5. (a) Exclusive (b) Inclusive
   (c) Inclusive (d) Exclusive
6. (a) T (b) F (c) T (d) T
7. (a) Let p be 'The barometer is falling'; q, 'It will rain'
   $\sim(p \to q)$
8. (a) If the zoo is well stocked with animals then entrance charges are high and attendances are large
9. F
10. T
11. F

Exercise 8a

The truth values are:
1. TFTT
2. FFTT
3. FFFF
4. TTFFTTTT
5. FFFTTTT

Exercise 8b

2. The truth table is:

p	q	$[(p \to q)$	$\leftrightarrow$	$(p \wedge \sim q)]$
T	T	T	F	F
T	F	F	F	T
F	T	T	F	F
F	F	T	F	F
			*	

Exercise 8e

1, 3, 4, 5 are tautologies

Exercise 8f

3. The statements whose truth sets are given are $(p \vee q) \wedge \sim r$ and $\sim[\sim(p \vee q) \vee r]$. The truth tables are:

p	q	r	$(p \vee q)$	$\wedge$	$\sim r$
T	T	T	T	F	F
T	T	F	T	T	T
T	F	T	T	F	F
T	F	F	T	T	T
F	T	T	T	F	F
F	T	F	T	T	T
F	F	T	F	F	F
F	F	F	F	F	T
				*	

p	q	r	$\sim$	$[\sim(p \vee q)$	$\vee$	$r]$
T	T	T	F	F	T	T
T	T	F	T	F	F	F
T	F	T	F	F	T	T
T	F	F	T	F	F	F
F	T	T	F	F	T	T
F	T	F	F	F	F	F
F	F	T	F	F	T	T
F	F	F	F	F	T	F
			*			

Exercises on Chapter 8

1. (a) T (b) F
3. $(P' \cup Q')' \cup (P \cap Q')$. False
10. (a) T (b) T

Exercises on Chapters 7 and 8

1. (a) F (b) T
2. (a) T (b) F (c) T
3. $p \wedge q \wedge \sim r$
4. Sufficient but not necessary
8. The statement is a tautology
9. (iii) (a) $p \downarrow \sim q$ (b) $\sim p \downarrow q$
   (c) $q \downarrow p$ (d) $p \downarrow q$
   It is commutative
10. No
11. No
12. No

Exercise 9a
(a) $q \to \sim p$ (b) $\sim q \to p$
(c) $q \to p$
(d) $\sim r \to \sim (p \wedge q)$
(e) $(q \vee r) \to \sim (p \vee q)$
(f) $\sim q \to \sim p$

Exercise 9c
1. (a) $p \to \sim q$ (b) $\sim p \to \sim q$
   (c) $\sim p \to q$
2. (a) $\sim p \vee \sim q$ (b) $p \vee q$
   (c) $\sim q \vee p$ (d) $p \vee \sim q$
3. They are all valid

Exercise 9e
1. $q \to p$ 2. $p \to q$ 3. $q \to p$
4. $p \to q$ 5. $p \to q$ 6. $p \to q$
7. $q \to p$ 8. $\sim p \leftrightarrow q$
9. $q \to p$ 10. $p \to \sim q$

Exercise 9f
Only 4 is invalid

Exercises on Chapter 9
The following are valid:
   1, 2, 4, 5, 6, 7, 9, 10

Exercise 10a
1. (a) $A \cup (A \cap B) = A$
   (b) $(A \cap B)' = A' \cup B'$
   (c) $(A \cup B) \cap (A' \cup C)$
       $= (A \cap C) \cup (A' \cap B)$
   (d) $(A \cap B) \cup A' \cup B' = \mathscr{E}$
2. (a) $(a+b)' = a'b'$
   (b) $(a+1)(a.0) = 0$
   (c) $a + ab = a$
   (d) $a.b'.b = 0$
3. (a) $(a+b).a'.b' = 0$
   (b) $(a+1) + a.0 = 1$
   (c) $(a+c)(a'+b)(b+c)$
       $= (a+c)(a'+b)$
   (d) $(a+b)(b+c)(c+a)$
       $= ab + bc + ca$

Exercises on Chapters 1-10
2. Necessary but not sufficient
6. (i) (a) $p \uparrow q$ (b) $\sim p \uparrow q$
      (c) $q \uparrow p$ (d) $\sim q \uparrow \sim p$
7. (b) Yes
9. $\sim q$
11. (a) The clock is not slow or we
        will miss the train
    (c) Do not exceed 20 km/h or I
        will report you to the police
    (e) Either a man is not indus-
        trious or he succeeds
12. Invalid
13. (a) $\dfrac{2^9.13!}{4!\,9!}$ (b) $\dfrac{2^3.13!}{10!\,3!}$
14. 91
15. 39
16. (b) A tautology
17. (a) $(n-1)!$ (b) $\dfrac{(4n)!}{(n!)^4}$
18. Yes Yes
19. Three

20.
A	1	2	3	4	6	12
B	12	6	4	3	2	1

21. Valid
22. (a) $\{1\}$ (b) $\{0, 1\}$) $\{0, 1, -\tfrac{1}{2}\}$,
    (d) $\{0, 1, -\tfrac{1}{2}, \sqrt{2}, -\sqrt{2}\}$
23. $n = 0$
25. (a) Yes (b) Yes
26. Necessary but not sufficient
27. The reports are inconsistent
29. $\{(x, y) \mid x = n+1, y = 6n-5,$
    $n \in Z\}$
30. (a) True
    (b) (i) False (ii) True
32. 240
33. $S = \{(2, 3), (2, 5), (2, 11), (2, 17),$
    $(2, 29), (2, 41), (2, 59), (2, 71),$
    $(2, 101), \ldots\}$
35. (a) 28 (b) $\tfrac{1}{2}n(n-3)$
36. (a) 21 (b) 9 (c) 73
    (d) $E \cap N$ (e) $N - E$
    (f) $(E - N) \cap (Y \cup U)'$
37. Yes No

INDEX